MW00534998

PRAISE FOR JEFF'S PREVIOUS BOOKS

Teaming with Microbes

Will help everyone create rich, nurturing, living soil—from organic gardening devotees to weekend gardeners.

—*Michigan Gardener*

Should be required reading for everyone lucky enough to own a piece of land.

—*Seattle Post-Intelligencer*

A breakthrough book for the field of organic gardening . . . No comprehensive horticultural library should be without it.

—*American Gardener*

This intense little book may well change the way you garden.

—*St. Louis Post-Dispatch*

Sure to gain that well-thumbed look that any good garden book acquires as it is referred to repeatedly over the years.

—*Pacific Horticulture*

Teaming With Nutrients

If you really want to get a sense of how little we know about how plants use fertilizers—and I count myself in this league—you should read Lowenfels' latest book, *Teaming With Nutrients,* which gets deep into the weeds, so to speak, of the microscopic architecture of plants and the biochemical processes at play . . . Suffice to say, Lowenfels is a man who knows his anions.

—*Washington Post*

Your awe of plants and the lives they lead will increase if you "just read it," and you'll never garden the same old way again.

—*Muskogee Phoenix*

Colorful illustrations, plentiful and readable diagrams, and a well-executed chapter structure make this an indispensable resource.

—*Publishers Weekly*

Offers a crash course in discovering soil structure, fertility, and microbial actions.

—*Palm Springs Desert Sun*

Lowenfels offers a deeper understanding of the major and minor plant nutrients and delivers the necessary science in a conversational style that most gardeners will appreciate.

—*Monterey County Herald*

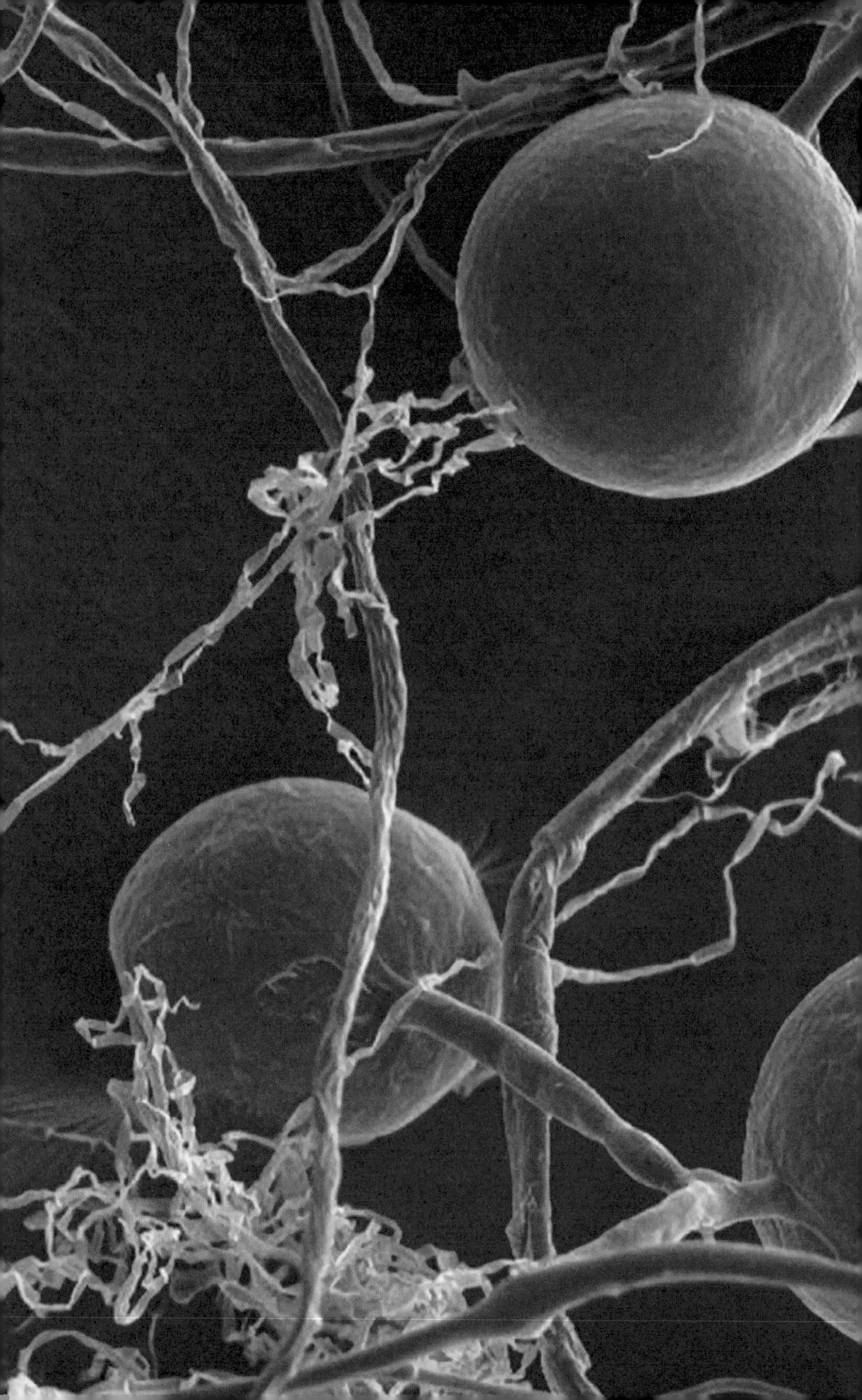

TEAMING WITH FUNGI

The Organic Grower's Guide to Mycorrhizae

JEFF LOWENFELS

TIMBER PRESS
Portland, Oregon

For Lisa, David, Madelyn, and Miles,
the next generation of gardeners.

Photo and illustration credits appear on page 160.

Published in 2017 by Timber Press, Inc.
The Haseltine Building
133 S.W. Second Avenue, Suite 450
Portland, Oregon 97204-3527
timberpress.com

Printed in China
Third printing 2019

Text design by Susan Applegate
Jacket design by Kristi Pfeffer (based on a series design by Susan Applegate)

Library of Congress Cataloging-in-Publication Data

Names: Lowenfels, Jeff, author.
Title: Teaming with fungi: the organic grower's
guide to mycorrhizae / Jeff Lowenfels.
Other titles: Organic grower's guide to mycorrhizae
Description: Portland, Oregon: Timber Press, 2017. | Includes
bibliographical references and index. | Description based on print
version record and CIP data provided by publisher; resource not viewed.
Identifiers: LCCN 2016021285 (print) | LCCN 2016017683 (ebook) |
ISBN 9781604697810 (e-book) | ISBN 9781604697292 (hardcover)
Subjects: LCSH: Mycorrhizas in agriculture. | Mycorrhizal fungi.
Classification: LCC SB106.M83 (print) | LCC SB106.M83 L69 2017 (ebook) |
DDC 631.4/6—dc23 LC record available at https://lccn.loc.gov/2016021285

A catalog record for this book is also available from the British Library.

CONTENTS

PREFACE

A STAGGERING 80 TO 95 percent of all terrestrial plants form symbiotic relationships with mycorrhizal fungi. In these relationships, or mycorrhizae (mycorrhiza, singular), the host plants supply the mycorrhizal fungi carbon, and in return, the fungi help roots obtain and absorb water and nutrients that the plants require. These relationships are vital to the health of almost all plants that grow on Earth. Each group of mycorrhizal fungi interacts and colonizes its plant host in a different way, in a process so complicated that it took scientists a long time to catch on to its importance.

If you are reading this book, you are probably familiar with the soil food web, the incredibly diverse community of organisms that inhabit the soil. Most of you understand the importance of symbiotic (mutually beneficial) relationships among plant roots and a multitude of soil organisms. You are aware of the relationships among bacteria, rhizobia, and legume plant roots that result in nitrogen fixation, and you understand that bacteria can form a symbiotic relationship with plant roots. Soil-borne mycorrhizal fungi, the subject of this book, interact with plant roots in a similar way.

Mycorrhizae have been known since 1885, when German scientist Albert Bernhard Frank compared pine trees grown in sterilized soil to those grown in soil inoculated with forest fungi. The seedlings in the inoculated soil grew faster and much larger than those in the sterilized soil. Nevertheless, not so long ago (in the 1990s), the importance of mycorrhizal fungi was unknown to many farmers and gardeners—and most garden writers. We feared and loathed all fungi, the stuff of mildews and wilts. Fungi were usually considered downright evil, and most of us took a one-size-fits-all fungicidal approach in our gardens. I had been writing a weekly garden column for some 25 years when I first heard the words *mycorrhizal* and *mycorrhizae* in 1995. I was embarrassed by my lack of knowledge of these important organisms, but when I asked my peers if they had ever heard of mycorrhizal fungi, they had no idea what I was talking about. (When I first started writing about mycorrhizae, not only did my word processor program's spell checker reject the word, but my editor did as well.)

WHY MYCORRHIZAE MATTER

Truth is, almost every plant in a garden, grown on a farm or orchard, or living in a forest, meadow, jungle, or desert forms a relationship with mycorrhizal fungi. In fact, many plants would probably not exist without their fungal partners.

It is now readily acknowledged that mycorrhizal fungi are important to agriculture, horticulture, silviculture (the practice of growing forests), and even hydroponics (in which plants are grown in water, without soil). Mycorrhizal fungi play a growing role in feeding the world. The benefits that plants derive from mycorrhizal fungi include increased uptake of nutrients, increased resistance to drought, increased resistance to root pathogens, earlier fruiting, and bigger fruits and plants. They are helping farmers better withstand drought and use less fertilizer, particularly phosphorus, because mycorrhizal fungi find and restore phosphorus in the soil, making it available to plant roots. Mycorrhizal fungi are also being used to remediate and reclaim spoiled land and to prevent erosion of valuable soils.

MYCORRHIZAL MYTHS

Mycorrhizal fungi are difficult to grow in a laboratory environment, and most are visible only with a microscope, so for a long time it was not easy for scientists to identify and study them. Before we understood much about mycorrhizal fungi (many mycologists originally thought they were pathogenic), only a small, dedicated group of scientists studied them. Even when they could be grown in a lab, these fungi didn't always behave as they would in an outdoor field study. It took a while to develop replication methods and to understand the conditions necessary for successful establishment of mycorrhizae. As a result, a lot of myths developed. This book aims to dispel many of these.

First, there is a presumption that mycorrhizal fungi are ubiquitous, so we do not need to add them to the soil. But considering that most of the soil around homes is not native, but has been brought in from somewhere else, it may not include the fungi your plants need. In addition, you may not have been treating your soil, and therefore its mycorrhizal fungi, in a way that benefits the fungi. Moreover, mycorrhizal fungi are competing with the native microorganisms in the soil, and populations of the fungi can be insufficient and far below those included in a commercial or homemade inoculum.

Next, some erroneously believe that although hundreds of native fungi can exist within soils, only a handful of these are available in commercial mixes, so they could not possibly be effective. Although these mixes contain only the most "promiscuous" of the mycorrhizal fungi, they are likely to colonize your plants even more so than any native fungi in your soil.

And, finally, there is the myth that plants inoculated with mycorrhizal fungi do not grow any larger or healthier than those to which no fungi have been applied. As you will learn in this book, this is often because the gardener makes mistakes, such as adding rich fertilizers that can disrupt or destroy the fungi, or otherwise mistreats the mycorrhizal fungi population. In addition, the benefits conferred by mycorrhizal fungi often have nothing to do with visible plant growth.

MYCORRHIZAL REALITIES

I learned gardening basics from my dad. He was an organic gardener, but the only fungi organic gardeners knew about back then were the bad ones. If you peruse the literature of the time, you won't find much information about fungi except for how to kill them. Even potting mixes were sterilized to kill fungi.

Dad grew orchids and had a big library of orchid books. None of them mentioned that orchids will not even germinate until they have a mycorrhizal support team on hand. If my dad were alive today or you asked a roomful of garden writers what mycorrhizal fungi are, he and every single one of them would now know.

There is simply no questioning the role of mycorrhizal fungi and the function of mycorrhizae in starting, growing, and sustaining plants. You can purchase mixes of mycorrhizal propagules from plant nurseries and big box stores. Soil mixes are advertised as containing mycorrhizal spores (the reproductive units that give rise to new individuals), fungal hyphae (branching cellular filaments), and mycorrhizal root fragments. Many commercial starter plants have been colonized (or infected) with mycorrhizal fungi. As interest in and use of mycorrhizal fungi grow, hundreds of new studies join the tens of thousands that already exist to increase our knowledge of how to grow and propagate mycorrhizal fungi, how they work with plants, and, most important for anyone who grows plants for a living or a hobby, when and how to use them.

A ROADMAP TO THE BOOK

There are two parts to this book: the first part concerns the science of fungi and mycorrhizae and the second covers application of the science.

You'll first be introduced to some general information about fungi—their appearance and their constituents, the different types of mycorrhizal fungi. Each type operates differently, and each of their characteristics and benefits is explored and explained. This book does not include much information about classification, however, which involves complicated descriptions of reproduction and reproductive bodies that may go beyond what most growers need to know.

You'll then learn about the uses of mycorrhizal fungi in agriculture, a discussion that is applicable to gardening as well, and about arbuscular mycorrhizal fungi that partner with annuals and crop plants.

You'll also explore the uses of mycorrhizal fungi in horticulture. If you have a nursery or greenhouse, this will be of particular interest.

Trees are mycorrhizal dependent, and the next discussion focuses on the associations with the same arbuscular mycorrhizal fungi used in agriculture and horticulture, along with a second type, the ectomycorrhizal fungi.

The next section covers the use of mycorrhizal fungi in hydroponics. You may not have expected that these fungi would be useful in a hydroponic growing environment, but they can be, and this discussion will dispel a few myco-myths.

For those who are interested in improving the quality of grasses growing in lawns and recreational fields, the next section will be pertinent. This is an area of increasing focus and commercial interest because of the environmental problems associated with the overuse of phosphorus and nitrogen in commercial fertilizers.

More and more companies are manufacturing mycorrhizal propagules. You'll learn about several ways you can grow your own.

There are always rules, including those that are applicable to the introduction, care, and maintenance of mycorrhizae. Farmer and gardener alike can take advantage of a lot of research to save some money by following these rules to ensure healthy mycorrhizae.

Finally (I couldn't help myself), the last section of the book includes a few thoughts about the future of mycorrhizal fungi and their uses. By the time you reach this part of the book, you'll have many more ideas to add.

At the end of the book are a few sources for obtaining mycorrhizal fungi and other products associated with growing them. You'll also find information about where you can learn more about mycorrhizae and keep abreast of new developments.

THE TRILOGY

This book is the third of a trilogy of books I've written about the workings of the soil food web and its organisms. These three books are related and fit together. You'll find lots of information in my two previous books, *Teaming with Microbes: The Organic Gardener's Guide to the Soil Food Web* (written with my good friend and business colleague, Wayne Lewis) and *Teaming with Nutrients: The Organic Gardener's Guide to Optimizing Plant Nutrition.*

Because these books are related, you are urged to read them all if you want to understand how plants, mycorrhizal fungi and other microbes, and nutrients interact in the soil—they are all interrelated parts of the soil food web. You cannot fully understand how mycorrhizal fungi operate in the soil unless you know about the soil food web. You cannot understand how phosphorus from the soil is transported to a leaf unless you know how plants take up nutrients. You cannot appreciate how a plant uses photosynthesis to make food for mycorrhizal fungi unless you combine soil food web knowledge and plant nutrient knowledge. It is all connected.

Mycorrhizal fungi are fascinating organisms that are much underappreciated by plant growers. Let's fix that.

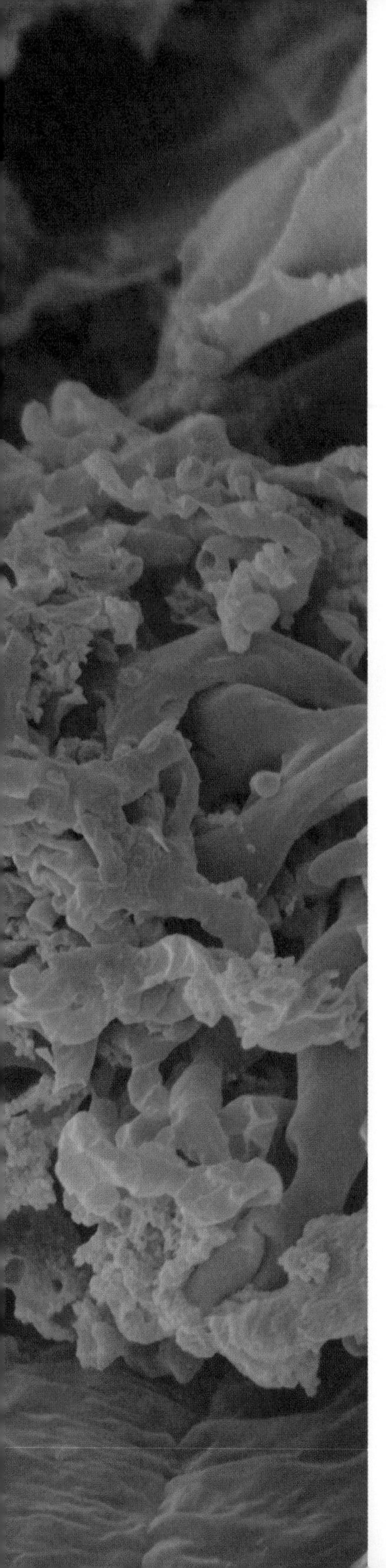

Fascinating Fungi

Most of the more than 100,000 species of known fungi live on land, but they are truly ubiquitous organisms, with some species capable of living in any ecosystem and clime—from deserts, to tundra, to polar caps, to oceans and other saline environments. Yeasts, for example, which are single-celled fungi, have been found on both of the Earth's polar ice caps. Lichens, which are formed by fungi and algae, grow throughout the Arctic as well as the temperate regions of the Earth. Fungi survive the extreme radiation of space. They exist throughout and cover your body (especially your feet). Some fungi appear to be specific to particular environments, such as the wood fungi that feast on the Antarctic huts built in the early 1900s by explorers Robert Scott and Ernest Shackleton, or the fungal mold that attacks the nopal cactus (*Opuntia* spp.) in the Mexican highlands.

Fungi play essential roles as decayers and nutrient recyclers (saprobes), parasites or pathogens, and symbiotic organisms (symbionts). Pathogenic fungi cause devastating crop losses and ravaging diseases in animals, ranging from ringworm to histoplasmosis. Fungi can cause death in humans, other animals, and plants. Many fungi can be beneficial, however, serving as biopesticides or contributing to the flavor of delicious foods such as cheese and all manner of intoxicating beverages. Other fungi have proven to be extremely important to human health: consider *Tolypocladium inflatum*, a fungus originally found in Norwegian soils, which produces cyclosporin, an immunosuppressant used in organ transplants to prevent rejection.

WHAT ARE FUNGI?

Fungi are fascinating organisms that sometimes resemble plants, sometimes resemble animals, and sometimes are uniquely fungal. Technically defined, fungi are eukaryotic organisms—that is, they consist of cells that contain membrane-bound nuclei, as do plants and animals. Scientists believe that fungi, plants, and animals once shared common ancestors, and all are grouped into the same kingdom: Eukaryota. Although they are related, these three types of organisms have enough distinct differences that each is placed in its own domain.

Not so long ago, in the 1950s, fungi were classified as plants. It is easy to understand why. Like plants, fungi have cell walls and many of the organelles present in plant cells: membrane-bound nuclei and vacuoles, ribosomes, mitochondria, and many others. In addition, fungi often live in soil, they are relatively immobile, and their morphology (structure) resembles that of plant roots and branches.

Unlike plants, however, fungi lack chlorophyll. They are heterotrophic, which means they are incapable of converting carbons into sugar to produce their own energy. They also lack the vascular systems found in animals and plants and can reproduce asexually through spores. Fungi cell walls are not full of cellulose like plant cell walls, and fungal walls contain the polysaccharide chitin, a main constituent in the exoskeletons of arthropods such as insects, lobsters, and crabs. In addition, fungi use glycogen, another polysaccharide, to store energy. This energy storage molecule is absent in plants, which store energy in starch molecules instead. Fungi also, at least initially, digest their food extracellularly—outside of their cells—which also differs from plants.

Fungi also share some common characteristics with animals. Both are heterotrophic, and both store their food as glycogen. Animals and fungi use enzymes to digest their food (as do plants, but not in the same way), though in animals, digestion occurs after ingestion. A fungus uses powerful enzymes to digest food outside of the organism before adsorption.

Of course, there are distinct differences between fungi and animals. Chief among them, animal membranes contain cholesterol, while fungal membranes contain a fungal-unique molecule, ergosterol, a steroid alcohol that converts to vitamin D when exposed to the sun's ultraviolet light. (This is why it is a good idea to expose edible mushrooms to the sun for a day or so before you eat them. And, by the way, a raw mushroom's chitin- and ergosterol-filled cell walls are nearly indigestible. Cooking mushrooms can not only avert a stomachache, but it helps release the nutrients locked inside their cells.) This alone is enough to separate the two organisms into separate domains.

FUNGAL COMPLEXITY

Fungi can be complex multicellular or simple unicellular (single cell) organisms. Mycorrhizal fungi are multicellular, comprising masses of branching filaments, or hyphae (hypha, singular). Hyphal fungi thrive in myriad environments, including soil if oxygen is available; they are aerobic organisms and need oxygen to survive. Hyphal fungi can be lurking on spoiled food in a container tucked away in the refrigerator.

The two forms of fungi: unicellular yeast (left) and multicellular hyphae (opposite).

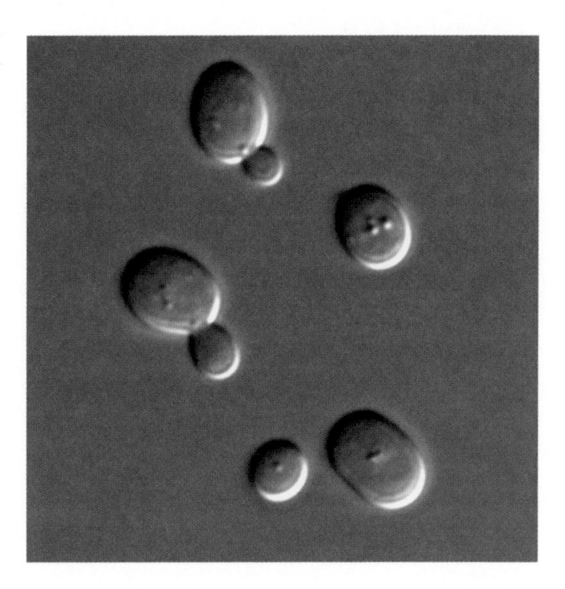

They grow in some cheeses such as brie and Roquefort. And they are the mildew that crops up on bread.

Yeasts are single-cell fungi—the type found in packets used to make bread and beer. Yeasts can survive in anaerobic (oxygen-free) conditions such as pockets in compacted soil; however, none of these types of unicellular fungi form mycorrhizae, so they are not important in this discussion and I will ignore them here. Instead, let's concentrate on the hyphal fungi, since all mycorrhizal fungi are of this type.

HYPHAL FUNGAL STRUCTURE

Each tiny mycorrhizal hyphal strand comprises connected cells, each filled with cytoplasm, organelles, and one or more nuclei. Tubular cell walls and septa (septum, singular) surround each cell and serve as structural supports. Septa are perforated by septal pores that allow cytoplasm to flow throughout a hypha's cellular compartments, transporting minerals, enzymes, and other intracellular materials between cells.

Most hyphal tubes have diameters of only 2 to 10 micrometers, though they can grow larger in some fungi. For comparison purposes, the diameter of a human hair is about 100 micrometers (1.0 micrometer equals 0.001 millimeter). Fungal hyphae can grow from a few centimeters to several meters in length. They are so thin that it takes hundreds of

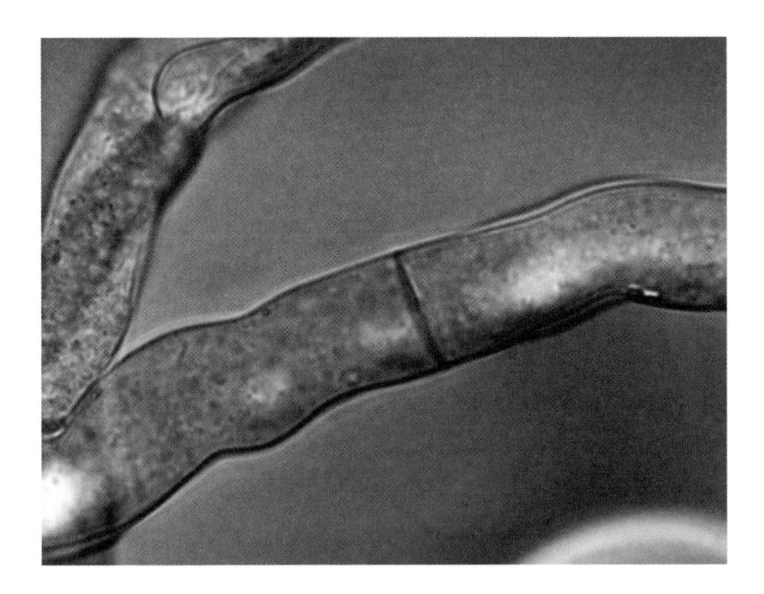

thousands of individual hyphal strands to form a network thick enough for the human eye to see. In fact, a single teaspoon of good garden soil may contain several meters of fungal hyphae that are invisible to the human eye.

Some fungi are not divided by septa. In nonseptate (also known as aseptate or coenocytic) hyphae, one big cell contains many nuclei. These hyphae do, however, usually form a type of septum at branching points and to wall off damaged areas.

Most of the organelles inside a fungal hypha cell are also found in plants and animals and they function in much the same ways. Chloroplasts, which are used for photosynthesis in plants, are not present in fungal cells, however.

The hyphal cell wall

Like plant cell walls, the outer cell walls of fungal hyphae protect the cells from changes in pressure and shield them from environmental stresses; they also contain some cellulose, a polysaccharide chain consisting of glucose, which makes up most of a plant's cell walls. Unlike plant cell walls, however, the rigid hyphal cell walls also contain chitin, a long-chain molecule (polymer). Chitin helps strengthen the cell walls and offers flexibility. (That is why you need special tools to crack

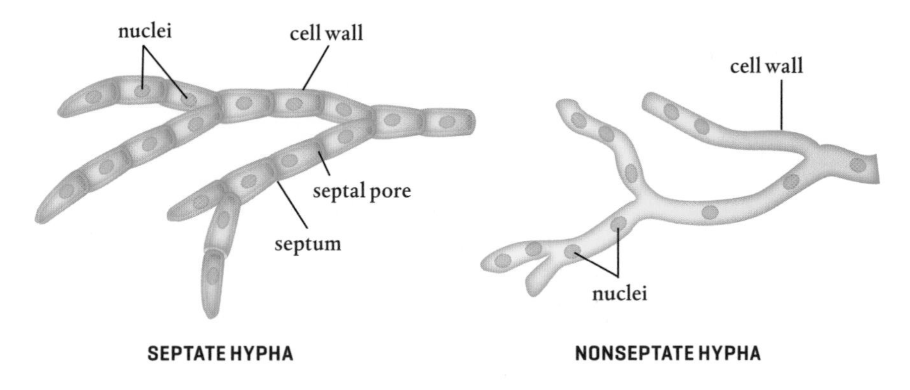

Hyphae are shaped like little trees, with branches that extend laterally. Septate hyphae are divided into compartments by septa.

GLOMALIN PRODUCTION

Some fungi in the phylum Glomeromycota—the arbuscular mycorrhizal fungi so important in agriculture and horticulture—produce glomalin, an essential glycoprotein that helps seal gaps and provides structural integrity to fungal cell walls. Sarah Wright, a soil research scientist for the U.S. Department of Agriculture (USDA), discovered glomalin in 1996. Before then, scientists couldn't identify this recalcitrant and long-lasting part of the soil.

It turns out that the mysterious soil ingredient comes from arbuscular mycorrhizal fungi. This large molecule includes not one, but two carbon sites (along with lots of nitrogen). Glomalin is an extremely stable molecule with great longevity in the soil. Its molecular structure is sticky and contributes to soil structure. As arbuscular fungi grow and die, they are continually adding glomalin to soils. In fact, glomalin contributes 27 to 30 percent of the carbon in soil where mycorrhizal fungi are present.

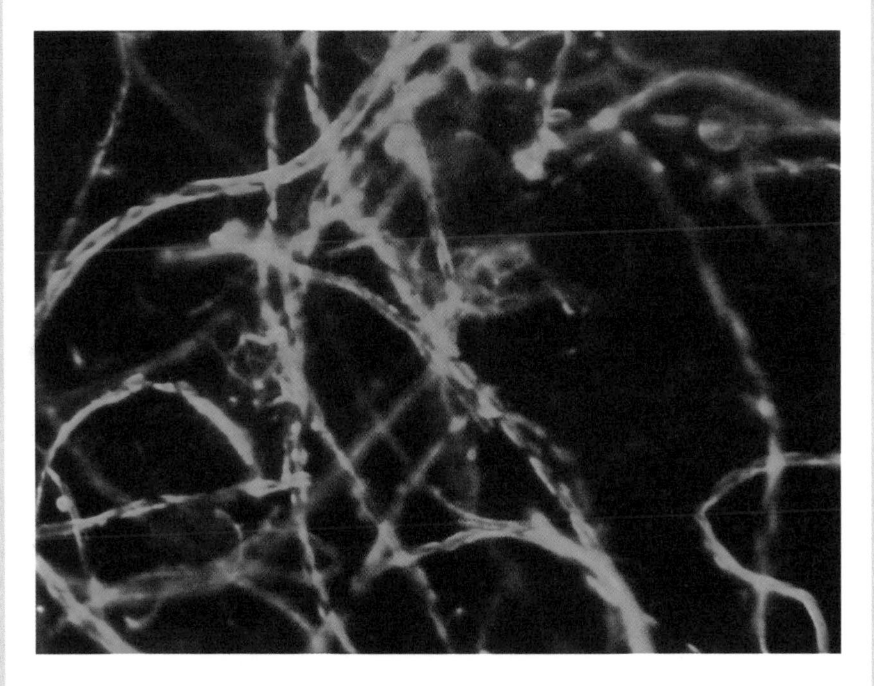

Glomalin, here stained green, contributes almost a third of the soil's carbon where mycorrhizal fungi are present.

open lobsters and crab legs—their chitin-rich shells are hard, but they also bend.)

Chitin is produced by enzymes in the plasma membrane layer just inside the cell wall and intertwined with fiberlike strands (microfibrils) set in a base of glycoproteins. Almost 20 percent of the wall can comprise these glycoproteins, carbohydrate chains that play structural roles and are important for nutrient uptake. While some glycoproteins have structural roles, others are used in signaling, sensing, and recognition. Still others make up the tunnels and channels located in the membrane and used for nutrient uptake. Fungal cell walls also contain ions of phosphorus, calcium, and magnesium, important nutrients that can enter the cell and travel to the growing tip of the hypha via microfilaments and microtubules.

Up to 80 percent of a fungal cell wall consists of polysaccharides, the long chains of sugars; fibrils made up of chitin and chitosan; and glucans. These are all mixed in a gel that forms a matrix. In addition, the pigment melanin, which protects the cell from damaging ultraviolet rays, is produced in the cell wall. The exact composition of the wall depends on the specific fungus as well as the time of life of the particular hypha.

As in plants and animals, the plasma membrane layer of a hyphal cell is nestled against the inner cell wall and envelops the cytoplasm and organelles. The plasma membrane is the molecular gatekeeper for

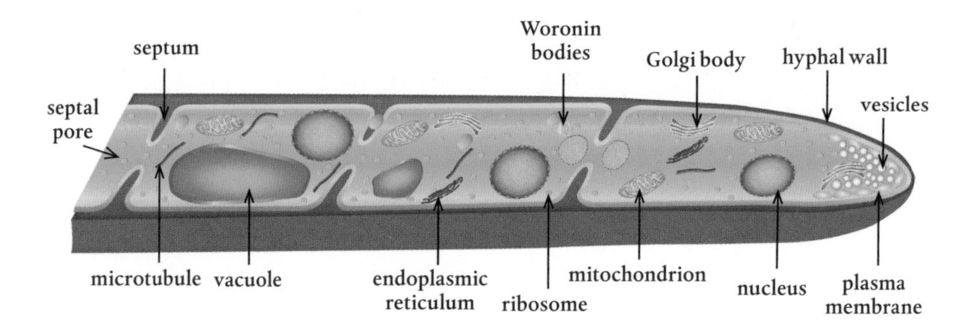

Fungal hypha cell with major organelles.

entry into the cell's cytoplasm. This layer is so thin it would take 5000 to 8000 layers to equal the width of a single page of paper used in this book!

Hydrophobins

Hydrophobins are often left off the list of differences between plants, animals, and fungi. These cysteine-rich proteins are present only in fungi and are always embedded in fungal cell walls. They form a coating on the cell's surface, which reduces water movement through the cell wall. Hydrophobins have both hydrophobic (or water repelling) sides and hydrophilic (or water attracting) sides. The hydrophilic side attaches to the fungal wall and keeps the cell from drying out. The hydrophobic side prevents water from rushing into the cell, which could dilute the cytoplasm and damage or drown the fungal hypha. The hydrophobic surfaces of these proteins also allow fungi to attach to other hyphae, bacteria, soil particles, plants, or fungal spores. These proteins are critical to the survival of hyphal fungi.

Monomers, the individual hydrophobin protein molecules, are assembled in the fungus and released from the growing tip of the hypha, where they stick to the outside of the cell wall. There they join together

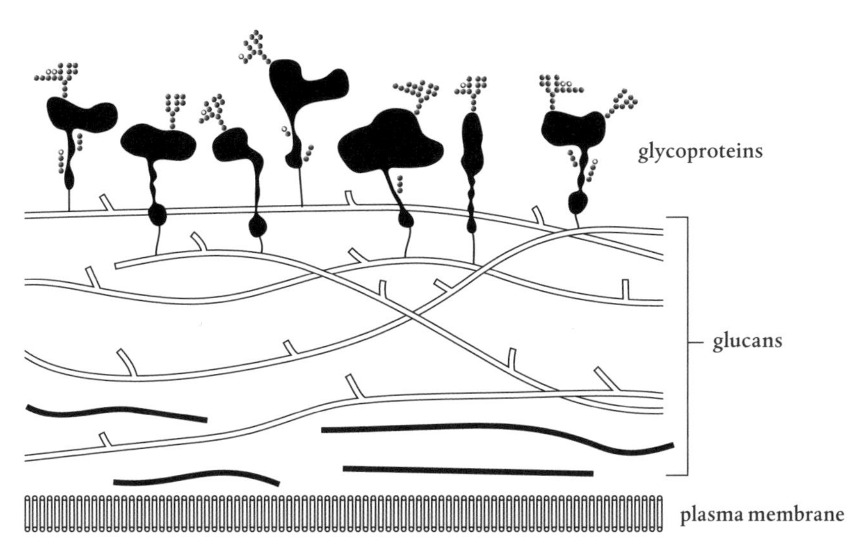

Cross section of a fungal cell wall and plasma membrane.

and form a solution that ensures that whenever a fungal hypha is out of the water, it is coated with molecules that keep it from drying out.

A fungus can contain ten or more different kinds of hydrophobins. Each has a specialized attribute that enables the fungus to function in diverse environments. Some help the fungal cell communicate inter- and intracellularly. Some control the flow to membranes, while others strengthen the hypha. These are complex molecules, and their full functions are under intense study because of their potential medical applications. Undoubtedly, there are agricultural and horticultural applications as well.

Septa

Septa divide a hypha into compartments. They can take several different structural forms, depending on the type of fungi. Some septa,

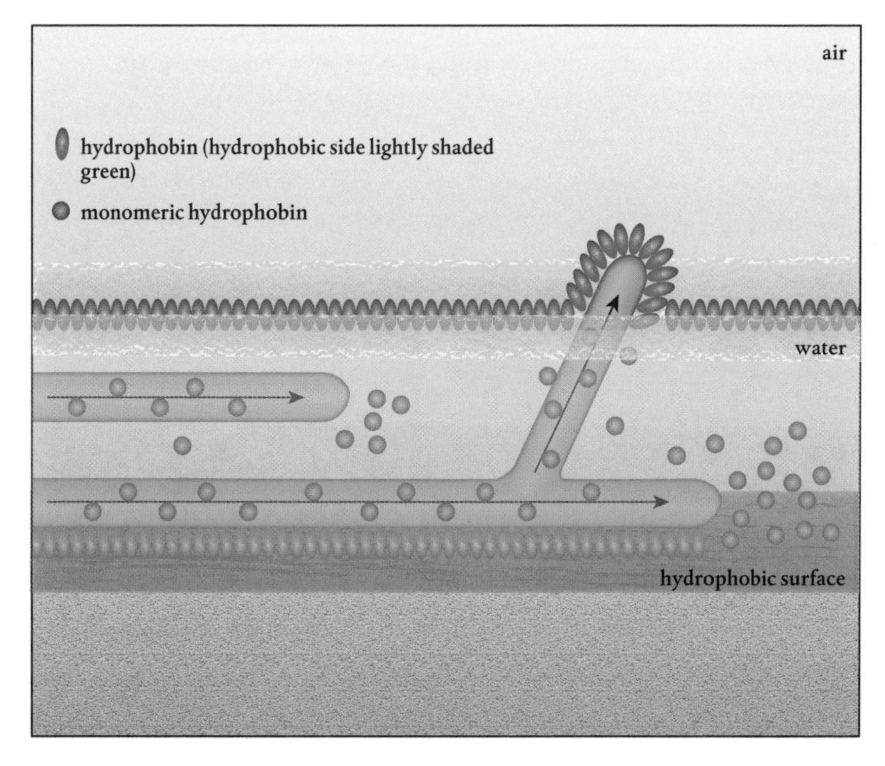

Individual hydrophobin molecules join together at the interfaces between water and air.

particularly those in cells of Ascomycota fungi, have one central pore of 50 to 500 nanometers in diameter, which is large enough for organelles to move through as cytoplasm flows. Basidiomycota fungi have special structures that prevent nuclei from moving through the septa but allow passage of cytoplasm and small organelles. Other fungi have sievelike septa that are riddled with pores—each septum can contain as many as 50 tiny pores, each 9 or 10 nanometers in diameter.

Septal pores located between the septa are often accompanied by Woronin bodies, special membrane-bound organelles that can fill up septal pores when a hypha is damaged and leaking cytoplasm, too old, or full of vacuoles with unwanted toxins. If Woronin bodies are absent, the cells can produce protein crystals that clog septal pores to seal off a section of a hypha when it is damaged or under attack.

All hyphal fungi, whether septate or nonseptate, are homokaryotic—that is, all the nuclei in the hyphal cells are genetically identical and share the same cytoplasm. One of the biggest differences between nonseptate and septate hyphae is that nutrients flow much quicker through nonseptate hyphae so nutrient uptake can occur faster than it does in septate hyphae. As a result, nonseptate hyphae can outcompete septate hyphae.

Plasma membrane

The plasma membrane nestled inside the fungal cell wall is similar in structure, function, and composition to the plasmalemma that sits against the inside of a plant or animal cell wall. It comprises a double layer of lipids, in which individual phospholipid molecules with hydrophilic heads and hydrophobic tails move freely (like the film on a soap bubble). In addition to phospholipid molecules, there are concentrations of sterols and sphingolipids (waxes and fats) that pack and move together. Although the basic composition of the fungal plasma membrane can vary, one thing is constant: ergosterol is always found in fungal cell membranes and cholesterol is found in animal cell membranes.

The plasma membrane is embedded with signaling molecules and with proteins that act as carriers for nutrient molecules, selectively allowing molecules to move in or out of the cell. In fungal cells, embedded proteins include chitin synthase and glucan synthase, enzymes that act as pumps and carriers to move water, nutrients, and debris across the membrane actively or via diffusion or osmosis, just as they do in plants and animals—again suggesting the similar ancestral origin between plants and fungi. There are differences in the proteins embedded in various fungi membranes, probably because of the various metabolic processes that require different materials.

Vacuoles and vesicles

Vacuoles are intracellular storage bubbles filled with water and powerful enzymes that help destroy and build molecules. They are bound by double-layered membranes that separate their contents from the cell's cytoplasm and allow vacuoles to merge with one another and with other cellular double membranes. These membranes resemble those of other eukaryotes in structure and function, suggesting that vacuoles were once independent, free-living organisms whose functions have been adapted. Vacuoles regulate the amount of water and the pH levels in the cell, they store and isolate harmful materials, and they allow nutrients and other substances to move in and out of the cell. More than one vacuole can exist in each fungal cell or compartment.

Vacuoles are formed by the fusion of multiple vesicles. Vesicles are similar to vacuoles, but they are smaller. Specialized vesicles, lysosomes, are used in lysis, the breakdown or death of cells. They form when smaller vesicles merge and serve to contain and neutralize harmful molecules that would damage or kill the fungi if released into the

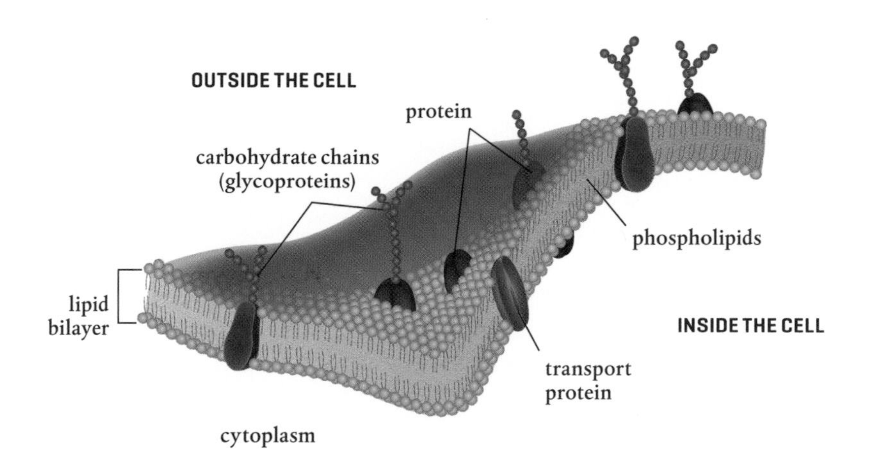

A portion of the fungal plasma membrane.

cellular cytoplasm. Peroxisomes are enzyme-filled vesicles used in breaking down long fatty acid chains, which is important for creating metabolic energy. They are also involved in breaking down amino acids and decomposing hydrogen peroxide (H_2O_2), a byproduct of digestion that is harmful to fungal organelles.

Nucleus

Each fungal cell contains one or more membraned nuclei. This is very different from plant and animal cells, which can have only one nucleus. In some fungi, free-floating nuclei can move through the septa between cells. These nuclei hold the genetic material, DNA, responsible for reproduction. Inside the nucleus is a nucleolus that produces ribosomes and contains proteins and RNA. The number of chromosomes in the cell, as well as the size and number of nuclei, depend on the individual fungus type.

Endoplasmic reticulum and ribosomes

The membrane surrounding the nucleus has large (for a fungus), tube-like extensions that create the endoplasmic reticulum. This tubular network transports proteins, synthesizes lipids, and stores calcium used in signaling and carbohydrate metabolism. It also produces new membrane material for the cell membranes, vacuoles, and vesicles, and it adds carbohydrates to glycoproteins.

As with plant and animal cells, the endoplasmic reticulum is divided into two parts—rough and smooth. The rough endoplasmic reticulum holds on to lots of ribosomes, protein-building machines, and also contain RNA and amino acids. Further along, as the endoplasmic reticulum becomes less populated by ribosomes, it smooths out; in this area, lipids are created to export or use in membrane creation.

Golgi bodies

Golgi bodies (named after Italian histologist Camillo Golgi), often called the Golgi apparatuses, are flat-membraned compartments that sort and package molecules for transport. They are stacked in plant and animal cells, but in some fungi this is not the case, which led some scientists to think that fungi did not have Golgi bodies. The electron microscope solved that mystery.

Golgi bodies finish up the jobs started in the endoplasmic reticulum. Specialized enzymes trim and cut up proteins and lipid chains that are synthesized in the endoplasmic reticulum to be routed inside or outside the cell. The modified molecules are loaded after labeling, so to speak, in the Golgi. The whole process is like the operations in a massive shipping hub that can pack and ship materials to ensure that they arrive at their destinations on time.

Mitochondria

Mitochondria (mitochondrion, singular) serve as little generators that supply energy throughout the fungal cells. These double-membrane–bound organelles are essential to the existence of all higher life forms on Earth. They create adenosine triphosphate (ATP), the source of energy used throughout living cells, and serve as the seat of aerobic respiration in the fungi. They also power the transport of material to the Golgi bodies, create intracellular membranes and associated embedded proteins, and control cell growth and death. Mitochondria produce proteins for their own use and contain their own DNA, suggesting that their ancestors were free-living organisms that entered into a mutual endosymbiotic relationship with eukaryotes.

Cytoskeleton and microtubules

The cytoskeletal system in a fungus supports the structure of the cell from the inside. It comprises a complex transportation network of fibrils, microtubules, and tubules. Molecular motors in microtubules transport materials, organelles, nuclei, and other important cargo throughout the fungal hyphae and are so important to how fungi function that you can kill a fungus by interfering with the synthesis of microtubules. The cytoskeleton's microtubules also play a vital role in cell growth. You may remember coming across diagrams of mitosis (cell division in which pairs of chromosomes split and are aligned) in high school biology. The movement of chromosomes during mitosis is also aided by microtubules, which are visible under a common microscope. In addition, the chitinous material made in the cytosol (the liquid part of the cytoplasm) is transported to the cell wall via microtubules, which are important to the growth of fungal hyphae tips.

MISSING IN ACTION

Notably missing from the mentioned fungal organelles are chloroplasts, which are found in plants. Fungi must have missed out on that bit of evolutionary symbiosis (which is why there are mycorrhizal fungi in the first place, I suppose).

Fungi lack vascular systems, the xylem tubes that transport water and phloem vessels that move food in plants. They also lack an enclosed circulatory system, which is present in animals. Some fungi, however, do form rhizomorphs, pseudovascular systems that resemble plant roots. These structures are created by fungal hyphae growing side by side and merging lengthwise to form stronger sheaths. They become devoid of cytoplasm and can serve as water and nutrient transport tubes, like tiny cellular aqueducts.

FUNGAL GROWTH AND DECAY

Hyphae can grow as much as 40 micrometers a minute. Cells grow apically, at the tips, in response to water, gases, and nutrients in the soil, moving toward favorable and away from unfavorable food sources. Hyphal growth gives a fungus the ability to move relatively long distances.

New molecules are constantly being pushed into the hyphal tip and along the cell wall, extending the wall outward and elongating the hyphal tube. The fungal cytoplasm transports vesicles containing the building materials necessary for wall construction. As the hyphae grow, they form new branches of hyphal tubes that explore and exploit new areas. Each tip of every newly formed hyphal branch performs exactly the same function as the original hypha.

Before electron microscopes, scientists using ordinary light microscopes identified dark spots in the tips of actively growing hyphae. When hyphal growth stopped, the spots disappeared. Since then, scientists have learned that the presence of this organelle, the Spitzenkörper, indicates not only an actively growing fungus but the direction of growth. All that a fungus is starts at the Spitzenkörper. The vesicles in the Spitzenkörper travel from the Golgi bodies through the microtubules and actin microfilaments. The protein actin is a major component of the fungal cytoskeleton and plays a crucial role in cell growth and intracellular transport.

Growing in an outward direction enables a hyphal tip to exert tremendous force. It has been reported that some fungal hyphae can produce 1200 pounds of force per square inch (218 kilos per centimeter), which is amazing when you consider that all this happens because of a few enzymatic processes.

While growth is occurring in the apical zone, at the other end of the hypha, in the older parts, things are deteriorating. The cells undergo autolysis, or self-destruction. What can be reused is moved toward the hyphal growing tip, while the nutrients contained in the rest of the decaying fungus become available to other organisms in the soil food web. In addition, hyphae leave behind a system of microscopic tunnels in the soil, through which air and water can flow, contributing to soil structure and health.

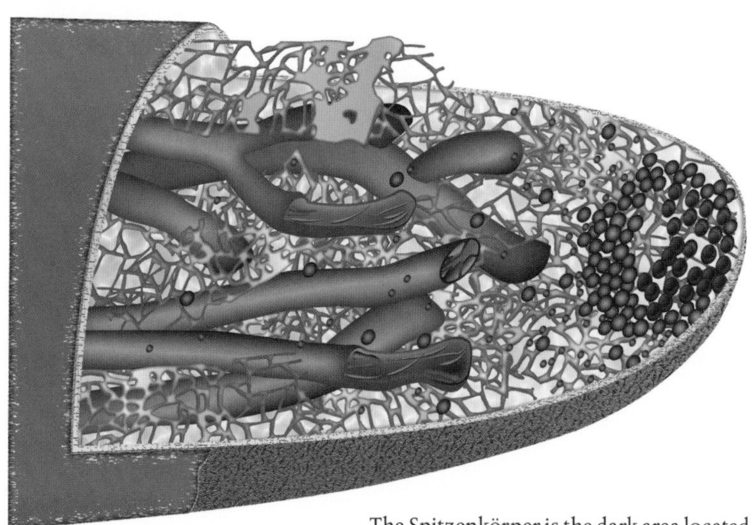

The Spitzenkörper is the dark area located at the hyphal tip. The microtubules (shown in red) and actin microfilaments (shown in brown) help transport vesicles produced in Golgi bodies to the tip.

HETEROTROPHIC CATEGORIES

Unlike plants, fungi are heterotrophic, incapable of making their own nutrients via photosynthesis. Like animals, fungi must obtain food from other organisms. Heterotrophs are categorized in several ways, based on how they derive their food.

Saprophytic fungi, or saprobes, are nature's recyclers. They absorb nutrients from dead or decaying organic material, such as fallen trees, dead insects and animals, and even animal wastes. These fungi release minerals for plant uptake and in the process play a key role in the operations of soil food webs. Without saprobes, forest organic material would accumulate ad infinitum.

Parasitic fungi obtain their food from the cells of living organisms, attacking and sometimes even killing their hosts; for example, the *Ophiocordyceps unilateralis* fungus infects and kills certain species of ants. They cause skin problems such as ringworm and athlete's foot and plant diseases such as apple scab, mildew, and rust.

Some facultative fungi can function as parasites or as saprobes, depending on the available food sources. Facultative saprobes can turn parasitic and invade a living host to obtain food. The opposite is also true: if a facultative parasite's host happens to die, the fungi will continue feeding as a saprobe. Unlike parasitic fungi, which usually don't kill their

A red-banded polypore, sometimes called a conk, is a saprobe fungus that absorbs nutrients from dead or decaying organic material.

host lest they endanger their own well-being, facultative fungi have less reason to care if their partner lives or dies, because they can change to take advantage of either situation.

Symbiont fungi are mutualistic, deriving food from hosts in a way that mutually benefits both organisms. Mutualistic fungi are partners in all lichens, which consist of a fungal symbiont and an algal or cyanobacterial partner.

Mycorrhizal fungi are clearly mutualistic. The fungal symbiont provides its botanical symbiont (host plant) with nutrients and receives carbon in return. Mycorrhizal fungi are also saprobes, capable of breaking

Spores from the parasitic fungus *Ophiocordyceps unilateralis* infect and kill ants (here *Camponotus leonardi*). The stoma of the fungus emerges from the back of the ant's head, and the perithecium, from which spores are produced, grows from one side of the stoma.

down organic materials as well as mineralizing inorganic matter. They are not the best decomposers of lignin and cellulose, but they can nibble at them. They are weak saprobes and decompose material that is already in an advanced stage of decay.

FUNGAL DIGESTION AND NUTRIENT ABSORPTION

An enormous range of food is available to fungi, and it seems that no matter what the material, a fungus has evolved to digest it (consider, for example, *Amorphotheca resinae*, the kerosene fungus, which eats creosote). Some generalist fungi can digest a wide range of materials, others have specific nutrient requirements, and many are in between. Many fungal species have developed the ability to break down and digest certain kinds of food, producing and releasing particular enzymes according to the environment in which they are usually found.

Although digestion is always extracellular, or outside the cell, fungi must internally produce the enzymes needed to break down their food sources. Because digestion occurs just outside the hyphal tip, most of the enzymes for digestion are released there after being assembled in the Spitzenkörper.

The vesicles in the constantly moving cytoplasm also include enzymes for activating growth and taking apart the cell walls at the hyphal tips, which is necessary so that new sections can be inserted. Several powerful enzymes capable of dissolving all but the most recalcitrant

Hyphae excrete enzymes that enhance the uptake of nutrients.

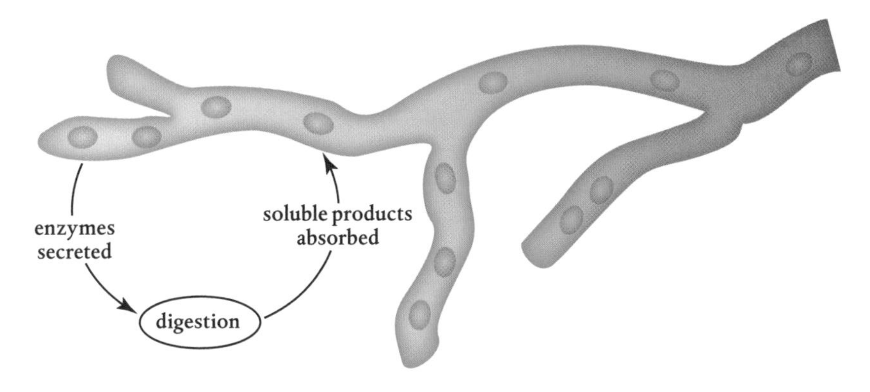

carbon compounds are released as the new cell walls are put in place. All this requires tremendous amounts of energy, which is ultimately supplied by the host or the fungal symbiont.

A range of isozymes, complex specialized enzymes, work together to break down foods. These powerful enzymes convert lignin, cellulose, and other tough organic matter into simple sugars and amino acids that are small enough to be absorbed through the plasma membrane in the hyphal tip. Each isozyme has some small variation in how it does its job, breaking down different parts of the material synergistically and rapidly to ensure complete decay.

Enzymes are released through the plasma membrane and are delivered by membrane-bound intracellular vesicles that merge with the membrane, spilling or releasing their contents into the cell. In this exocytosis process, some of these enzymes work directly within the cell wall while others are diffused through to the extracellular environment. Many of these remain in the soil as the hyphae move on, breaking down organic matter and making it available to plants and animals in the soil food web.

Two biological processes move oxygen, water, and nutrients across

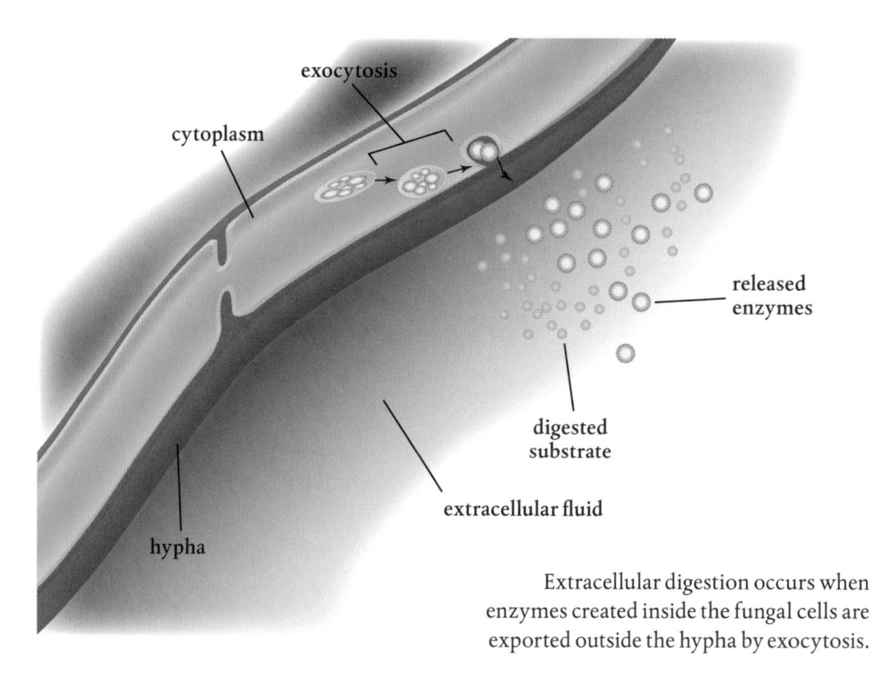

Extracellular digestion occurs when enzymes created inside the fungal cells are exported outside the hypha by exocytosis.

the cell wall and plasma membrane, into and out of the cell. In passive transport, molecules pass directly through the plasma membrane from areas of higher concentration to areas of lower concentration to maintain equilibrium inside and outside the cell. This kind of transport involves diffusion and osmosis. Active transport, on the other hand, uses a chemical energy source, ATP, the high-energy molecule found in every cell, to pump larger molecules against the concentration gradient and through the plasma membrane.

The fungal plasma membrane is quite the gatekeeper. The molecules absorbed passively through the membrane are very small, about 4000 to 6000 Daltons in size. (Abbreviated Da, 1 Dalton is approximately the mass of a single proton or neutron. For comparison, an average DNA strand is huge, at about 900,000 Da.) Only water, oxygen, carbon dioxide, and very simple sugar molecules are small enough to make the grade. Molecules too large to be absorbed use specialized transport proteins made by the fungus and embedded within the plasma membrane that control what enters and exits the cell. These proteins operate exactly like the embedded protein transporters in the plasmalemma in plants and animals. Each protein is constructed within the cell and then embedded into the membrane. With very few exceptions, only charged ion particles pass through the channels formed by these proteins. Each embedded protein is monoparticular, allowing only one specific kind of molecule to pass through—so, for example, phosphate ions cannot move through an embedded protein designed for calcium ions. Each to its own.

Critical nutrients passed by transport proteins

As with plants and animals, each of the several nutrient molecules, such as potassium and phosphorus, requires a different carrier protein to transport it across the plasma membrane into the fungal cell. These nutrient molecules attach themselves to the carrier, which ferries them across the membrane via carrier-facilitated active transport. Very often, larger sugars enter fungal cells in the same way, using specific carrier proteins. Sugars are critical to the heterotrophic fungi and enter the fungal cytoplasm both passively and actively, emphasizing their importance as nutrients.

Water too is critical to all life, and without the ability to absorb some water, a fungus, like any living organism, would die. Because of water's small molecular size and bipolar nature (with positive and negative charges on opposite ends of each molecule), water molecules can enter a fungal cell directly though the fungal membrane by diffusion. But if a fungal cell needs more water, it brings it in through aquaporin proteins that form little channels through the plasma membrane.

Nitrogen molecules, which build proteins and nucleic acids, are derived by fungi from the decay of organic and inorganic sources. Without nitrogen, there would be no enzymes. The molecules enter the cell as nitrates (NO_3-) and ammonium (NH_4+) via diffusion. In some fungi, in order to take up nitrates, glucose must be present in the cytoplasm.

Celluloses are digested by the cellulase enzyme, which results in glucose and other sugars. These molecules require active transport carrier proteins to help move them across the plasma membrane. Some fungal cells use more than one carrier system to transport the same kind of sugar; this stresses the importance of sugar to fungal organisms that cannot create their own because they do not photosynthesize.

Finally, amino acids freed up by extracellular digestion are also needed to create proteins. These large molecules require active transport via a carrier protein to move into a fungal cell. Clearly, extracellular digestion and the absorption of nutrients into the fungi is a complicated dance worthy of its own book.

Secondary metabolites

In addition to releasing extracellular digestive enzymes, fungi produce and release secondary metabolites, which are organic chemicals unrelated to digestion. Some of these products control or impact other microbes in the vicinity, some speed up or slow down the growth of host plants, and others can kill other plants' roots. These metabolites include well-known antibiotics such as penicillin, while others, such as aflatoxins, are deadly to humans and other animals.

Metabolites produced by mycorrhizal fungi are used in the mycorrhizosphere, the area surrounding the fungal hypha, which corresponds to the rhizosphere, the area immediately surrounding a plant root. These metabolites act just like the exudates from plant roots in the rhizosphere by attracting bacteria, other fungi, and microorganisms. Some of the benefits of mycorrhizae probably result from these organisms. Fungi can adjust their metabolites to attract, sustain, or repel them. One group of organisms in particular, phosphorus-solubilizing bacteria, are known to help with the digestion of phosphorus.

FUNGAL REPRODUCTION

Fungi are often placed into one of two categories: perfect fungi are capable of making spores and can reproduce either asexually (with only one parent) or sexually (with two parents). Imperfect fungi reproduce only by asexual spores, or at least they have never been seen to reproduce sexually.

The most common type of fungal asexual reproduction is through the formation of spores that are dispersed from the parent organism. Some fungi develop sporangia, budlike cells that hold onto young sporangiospores until they mature and separate from the parent cell. Other fungi develop spores outside the reproductive cells. Still others create sclerotia, storage structures made up of compact fungal tissue, some of which can survive dormancy for long periods in adverse environmental conditions.

Fungi can also be asexually cloned from hyphal cuttings. Some

Large, arbuscular mycorrhizal fungal spores.

mycorrhizal fungi form vesicles that can reproduce or clone the fungi. Still, nothing beats sporulation in terms of numbers. It is the reason fungi are so prolific and successful.

Sexual reproduction requires two mating types and can occur in several ways. In some fungal species, meiosis (nuclear division) occurs in the asci, single-celled spore structures, which results in the formation of exactly eight spores. These fungi shoot out their spores to distances of up to several centimeters. Given the size of the organism, this is comparable to tossing an American football a couple of miles (or kilometers). Consider the osmotic pressure that must be involved. In other species, basidia structures develop to produce and hold or eject spores. The pressure in this case is decidedly weaker. These tiny spores are dispersed by wind currents and can travel long distances.

One other point is important to mention. Scientists have discovered that fungi exchange DNA with other fungi and with other microorganisms. (Don't be surprised, because human bacteria do it all the time.) They seem to do this based on environmental conditions that require some sort of adaptation. This has all manner of ramifications, many of which are being sorted out by lots of further study.

MUSHROOMS, THE FRUITING BODIES

Some mycorrhizal fungi types (specifically, ectomycorrhizal fungi) form mushrooms, the specialized fruiting bodies that play a role similar to that of flowers in plants. Mushrooms can grow below the soil, as do truffles, or above, as do morels. Their caps can be gilled or nongilled. Some have economic and gastronomic value, such as porcinis and chanterelles. Not all mycorrhizal fungi produce mushrooms, although all mushrooms are fungi.

Mushrooms are typically the part of the fungus that carries and disperses fungal spores, usually through sporangia (sporangium, singular), the enclosures in which the spores are formed. Each tiny spore released from a sporangium gives rise to a hypha, which branches and grows into hyphae, which then continue to grow and interweave to form a mycelial web.

Sometimes the hyphae form rhizomorphs, structures that transport water and nutrients throughout the network. This occurs when parallel hyphae merge. The inner hyphae lose their nuclei and cytoplasm, becoming hollow.

Merging hyphae also form fruiting bodies as a result of hyphae growing together. Specifically, two homokaryotic fungal hyphae (each with genetically identical nuclei) merge cytoplasm, but keep their nuclei separate, to form a dikaryotic hypha. Every time the cell divides, the new cells keep the two sets of haploid nuclei separate.

At the appropriate time, given some cue from nature, some environmental condition stimulates the fungus, and signals are created in the circulating cytoplasm, which cause the hyphae to merge and form a primordium, a tiny knot in the mycelia. This primordium includes all the cells the mushroom will ever have. It enlarges into a tiny budlike structure, or button, which will grow into a full-size mushroom.

All a mushroom requires is water to fill the cells. The mushroom can rapidly pull in water from its mycelium and expand the primordium. It

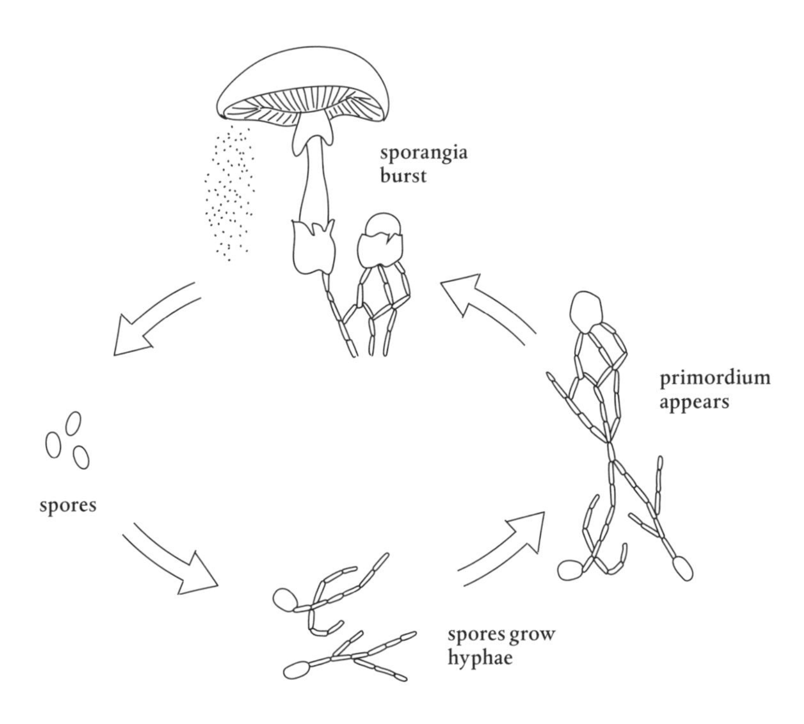

How mushrooms are formed.

THE MYCELIUM AND HYPHAL GROWTH

The mycelium is the network of branched hyphae filaments that grow out into the soil. The mycelium absorbs nutrients from the environment by secreting enzymes into a food source to break down polymers into monomers, which are absorbed into the mycelium. A vital part of the decomposition of plant material, the mycelium contributes to the organic materials in the soil and releases carbon dioxide into the atmosphere. The mycelium of mycorrhizal fungi increases the efficiency of water and nutrient absorption of host plants and helps provide resistance to pathogens.

Mycelia merge together to form networks of webs, or colonies, that can vary from microscopic in size to massive, three-dimensional organisms that radiate in all directions. Mycelia may be too small to see, or they may be extensive. In fact, the size of an arbuscular mycorrhizal fungal network can be staggering. A teaspoon of soil from a rye plant, for example, can contain up to 3 miles (about 5 kilometers) of mycorrhizal fungi. It's easy to see how this mycelial network would create tremendous surface area for nutrient and water absorption.

As long as there is food and water, the hyphae continue to grow outward. How far? Until the 1990s, fairy rings, circles of mushrooms sometimes observed in lawns, were thought to be the largest examples of mycelial growth. In 2000, however, a mycelium of *Armillaria ostoyae* was discovered in soil in Oregon's Blue Mountains. This humongous fungus covers 2384 acres (965 hectares) and is estimated to be more than 2400 years old. Some of the oldest known fungi, specimens of *A. ostoyae* are among the largest living organisms on Earth.

Colonies of hypha can grow rapidly depending on environmental conditions as well as the type of fungi. A single hypha of *Neurospora crassa*, for example, can add up to 40 micrometers of new growth per minute. (Imagine waking up and finding yourself covered in fungal threads, à la Jonathan Swift's Gulliver.)

It may be difficult to imagine that mushrooms are made up of mycelia, but they certainly are. The incredible number of different types of mushrooms indicates the size and diversity of mycelial fungi.

can take less than 24 hours for a primordium to fill and grow; this is why mushrooms can pop up so quickly. Mushrooms have outer membranes, or veils, that dehydrate rapidly, so they usually appear only when temperature and moisture conditions are right. Those hydrophobins are working full time.

FUNGAL CLASSIFICATION

The formal classification of fungi, whether perfect fungi or imperfect fungi, is a complicated subject and an ever-changing landscape, as new fungi are discovered and new classification systems developed. No one knows exactly how many different species of fungi exist, but about 100,000 have been identified and studied thus far. Scientists using modeling methods have estimated that as many as 5.1 million fungal species may exist (not surprising, considering that nearly 200 different species of fungi live on your feet). Fungal diversity is huge, to say the least, and scientists have developed a number of ways to subdivide and otherwise classify them.

Science has determined a system of classification of all living things, a hierarchical order based on the genetic characteristics of organisms. Fungi are classified as members of the Eukaryote domain at the top of the order. Next in line, the kingdom Fungi includes heterotrophic organisms that obtain food through absorption. Until the late 1960s, fungi were considered part of the plant kingdom (Plantae), but they were given their own kingdom when scientists finally concluded that the differences between plants and fungi were too great. The kingdom Fungi is now recognized as one of the oldest and largest groups of living organisms on the planet.

The fungal phylum groupings have been based primarily on the types of spores produced for reproduction and the reproductive structures that create them. DNA sequencing, however, is suddenly ending a lot of debate about what is a fungus and what is not and making it easier to make phylum assignments. In fact, all that is needed to determine a fungus's phyla is some of its DNA. Even better, RNA can be used to identify individual species. Most scientists now recognize seven phyla of fungi, but only three of these include mycorrhizal fungi.

THE NAME GAME

Nothing is simple when it comes to mycology, the study of fungi, especially regarding classification and naming. For example, arbuscular mycorrhizal fungi were all included in the phylum Zygomycota until 2001, when, after the discovery of some unique molecular, morphological, and ecological characteristics, they were moved to Glomeromycota. This complicated things then, and it is still confusing.

Because of reclassification, a popular commercial arbuscular fungus, *Glomus intraradices*, is now known as *Rhizophagus intraradices*. This fungus is the focus of many of the studies on arbuscular mycorrhizal fungi; it colonizes rice, corn, soybean, wheat, alfalfa, cannabis, and hemp—in fact, nearly all important agricultural crops. Commercial labels are slow to catch up to science, and some still use the older name.

All past research documents are not going to be corrected to use the new nomenclature. It takes time to change, so, reluctantly and to avoid confusion, I will stick with the vernacular and occasionally remind the reader that change is coming. In the meantime, look for either *Glomus intraradices* or *Rhizophagus intraradices* (or even *R. irregularis*) propagules on package labels (until the name changes again).

Ascomycota

Ascomycota, also called sac fungi, is a diverse and large phylum that includes more than 30,000 species. All ascomycetes develop an ascus, a saclike reproductive structure that contains tiny ascospores. Many ascomycetes are plant or animal pathogens, and others are edible mushrooms. It is the phylum to which morels, truffles, brewer's and baker's yeasts, and penicillin belong. Less helpful ascomycetes include powdery mildews and the fungi responsible for apple scab, Dutch elm disease, and chestnut blight. These fungi break down cellulose in plants and collagen in animals, two extremely resistant materials.

Basidiomycota

Members of this phylum of about 30,000 filamentous fungi reproduce via a microscopic, spore-producing basidium, which is attached to a fruiting body. The word *basidium* means little pedestal and is derived from the way the fruiting body supports the spores. This phylum includes mushrooms, chanterelles, puffballs, stinkhorns, bird's nest fungi, shelf fungi, rusts, and smuts.

Glomeromycota

The phylum Glomeromycota includes fewer species than the other two phyla, at about 230, but they are among the most abundant and widespread of all fungi. Arbuscular mycorrhizal fungi are glomeromycetes that form mutualistic relationships with the roots of most herbaceous plants and many trees. Most species reproduce asexually. These fungi play a large role in this book's story.

The contrary webcap, *Cortinarius varius* (left), is a basidiomycete, as is the common earthball, *Scleroderma citrinum* (below).

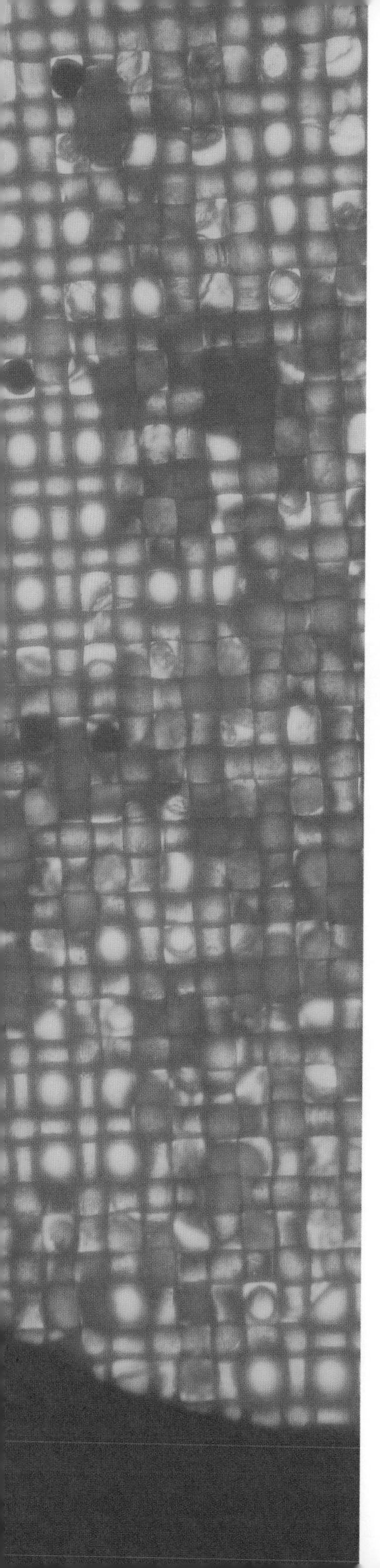

Introduction to Mycorrhizal Fungi

MYCORRHIZAL FUNGI HAVE existed and supported plants since terrestrial plants evolved more than 450 million years ago. In fact, mycorrhizal associations are believed to be a major factor that enabled plants to survive on land. The earliest fossil records of plant roots contain arbuscular mycorrhizae that look almost identical to arbuscular mycorrhizae growing in modern soils.

Today we know the true importance of mycorrhizal relationships, and we understand how they operate: the fungus colonizes the root system of a host plant, increasing the roots' water and nutrient absorption capabilities, while the plant provides the fungus with carbon it obtains from photosynthesis. Both organisms thrive as a result of this symbiotic relationship.

At about 2 to 10 micrometers in diameter, mycorrhizal fungi hyphae are considerably thinner than root hairs, which average 15 micrometers. Some of the mycorrhizal hyphae are extraradical (or extramatrical), meaning they extend beyond the root into tiny soil pores that

WHAT'S IN A NAME?

The words *mycorrhizal*, *mycorrhiza*, and *mycorrhizae* are often used improperly. Many product labels, and even some published research, use the terms interchangeably and incorrectly. Here's how to use the terms correctly:

- *Mycorrhizal* is the adjective used to describe the fungus or fungi, as in mycorrhizal fungi. The term has nothing to do with roots or the relationship of fungus to root.
- *Mycorrhiza* is the singular noun used to describe the partnership of root and mycorrhizal fungus. The term always refers to the root and the fungus together.
- *Mycorrhizae* is the plural form of mycorrhiza. The term refers to relationships, associations, partnerships—between and consisting of both fungi and roots.

root hairs would be too large to penetrate. Mycorrhizal fungal hyphae can grow to several centimeters in length, which enables them to access nutrients and water that the roots cannot reach.

Although we now know that mycorrhizal fungi live on the roots of 80 to 95 percent of all Earth's plant species, not long ago little was known about these specialized fungi, except that they existed. Very few appreciated the crucial role these fungi play, and many even questioned whether they had any benefit at all or were detrimental to plants.

DISCOVERY AND GROWTH OF UNDERSTANDING

In 1885, German botanist Albert Bernhard Frank (1839–1900) published an extraordinary paper about what would later become known as ectomycorrhizal fungi. It would take more than 50 years for Frank's observations and analysis to be confirmed, especially his conclusion: that a beneficial, not a parasitic, relationship existed between the fungi and host plants. During that time, while the botanical community argued over this point, a thriving commercial fungicidal industry developed. Frank's fungi were deemed to have little importance.

In his paper, whose title in English roughly translates to "On the nutritional dependence of certain trees on root symbiosis with belowground fungi," Frank combined the Greek words for fungus (*mykitas*) and root (*riza*) to form the German word *mykorhizen*. He theorized that each partner provided nutrients to the other in a beneficial, not a detrimental, manner and that the fungi were not killing the trees he studied but supporting them, and in turn were being supported.

Frank also believed that the netlike structures that appeared around the roots of pines, as found by Theodor Hartig in 1840 (who believed that these were parasitic fungi), were in fact formed by the same mycorrhizal fungi, and that these fungi absorbed minerals from the soil and humus and brought them back to the plant in return for carbon. Previous observers had concluded that the netlike structures were fungal, but they all insisted that it was an invading fungus and not beneficial to the plant.

After continuing his research, Frank identified two types of mycorrhizal fungi: he called them endotrophic (in which the fungus penetrates the roots of the plant) and ectotrophic (in which the fungus is formed between the roots). One of his graduate students, Albert Schlicht, found both kinds of mycorrhizal fungi in a variety of environments throughout Germany, which led Frank to believe that mycorrhizae were the norm, not the exception. How right he proved to be.

Although Frank is often given credit for discovering mycorrhizae, endomycorrhizal fungi had been observed at least as early as 1842 by Swiss botanist Carl Wilhelm von Nägeli, though he may not have known exactly what he was looking at under his microscope. (You may not have heard of von Nägeli because he discouraged Gregor Mendel from continuing his genetic work with peas.)

For a long time after Frank's paper was published, scientists believed that mycorrhizal fungi were parasites. They were lumped together into a group called Rhizophagus, for root-eater. (Although this name did not endear these fungi to plant growers, it has returned to use as a genus of mycorrhizal fungi.) In the mid-1900s, scientists believed endomycorrhizal fungi were the cause of diseases that affected strawberries, tobacco plants, and legumes. Because the treelike structures, or arbuscules, inside the host plant roots are digested by the plant, scientists believed that the plant was protecting itself from an invading pathogen.

Research continued, and in 1943, during World War II, a Japanese scientist published a paper that showed that plants that formed mycorrhizae grew faster than nonmycorrhizal plants. The paper was written in a nearly unintelligible form of German and released at an unfortunate time, however, so few, if any, researchers were aware of his work. It wasn't until the mid-1950s, when a paper by scientist Barbara Mosse discussed improved apple seedling growth resulting from mycorrhizal fungi, that the true benefits of these fungi became clear.

WHO NEEDS MYCORRHIZAE?

Without mycorrhizal relationships, most plants would probably not exist. Most members of the plant kingdom have formed associations with mycorrhizal fungi, including bryophytes (such as mosses), angiosperms (most land plants), pteridophytes (such as ferns and club mosses), and many gymnosperms (such as conifers). These plant–fungal associations have remained pretty much the same despite all manner of other evolutionary changes that have occurred. Although mycorrhizae are important to most plants, not all plants depend on these relationships to the same degree. Some plants will not survive without mycorrhizal fungi, while a few others do not require mycorrhizae at all.

Obligatory mycorrhizal plants cannot survive without mycorrhizal associations. These plants often have thick, nonbranching roots that cannot effectively explore soil for nutrients. Their seeds may not germinate without mycorrhizal colonization, or if they germinate, seedlings will not survive.

Facultative mycorrhizal plants can survive without forming mycorrhizae until nutrients become scarce. When this happens, mycorrhizal associations form or the number of existing mycorrhizae increases. Roots of facultative mycorrhizal plants tend to be extensive and branching, and they are capable of reaching large volumes of soil and nutrients even without an extraradical mycorrhizal network.

BENEFITS OF MYCORRHIZAL FUNGI

Plants involved in a mycorrhizal relationship are better able to withstand drought and other environmental stresses, root pathogens, and even foliar diseases. Plant biomass is often improved, along with the timing and number of flowers and amount of fruit produced. Mycorrhizae also offer much to the soil by improving structure as the fungal hyphae explore for nutrients.

Mycorrhizae are sensitive to environmental conditions and can behave one way in a study under greenhouse conditions and another way outdoors in the field. To complicate the situation, mycorrhizal fungi can exchange DNA with other microorganisms and shift the benefits they confer. Nevertheless, much scientific research has confirmed the importance of mycorrhizae and the benefits of mycorrhizal associations to plants. Keep these benefits in mind as you read about the various types of and uses for mycorrhizal fungi.

Increased nutrient supply

As a root grows, it slowly takes up nutrients from the surrounding soil. The plant eventually uses up the available nutrients in this zone of depletion or takes them up at a rate that cannot be sustained at a high enough level for the plant to thrive. As a result, roots continue to grow to reach new areas of soil to mine for nutrients.

By associating with fungi, however, a plant's roots can use the nutrients brought back by the fungal hyphae growing beyond the zone of depletion. Moreover, because the hyphae are considerably thinner than the roots with which they associate, they are able to reach smaller pores in the soil within and outside the depletion zone that were not available

White masses of fungal hyphae surround the plant roots. Because they are so much smaller than roots, fungal hyphae can enter smaller soil pore spaces to reach more nutrients.

to the larger roots. Although the mycorrhizal hyphae are smaller than roots, they greatly add to the surface area and increase absorption. One study, for example, found that fungal hyphae associated with a pine tree added 20 percent more mass to the tree's root system.

Mycorrhizal fungi transfer nutrients to the plant. In return, around 20 percent of the carbon produced by a plant can be transferred to its fungal partner. That is some sacrifice, but the returns are worth it. Phosphorus, an essential plant nutrient especially important because of its presence in ATP, the energy currency of life, is tied up chemically in the soil and difficult for most plants to absorb without the help of mycorrhizal fungi. Nitrogen, copper, zinc, iron, and nickel (all essential plant nutrients) are also taken up by mycorrhizal fungi and transferred to the host plant.

Mycorrhizal hyphae absorb the nutrients already in the soil, but their extracellular digestion can also result in decay that releases even more nutrients into the soil, especially nitrogen. In addition, as organisms in the mycorrhizosphere die, they release a variety of nutrients and produce all manner of metabolites that are absorbed by the fungi. And the fungi themselves are a food source for other soil food web organisms.

Drought tolerance

The existence of mycorrhizae can help host plants be more tolerant of drought conditions. Studies have demonstrated that from 50 to 100

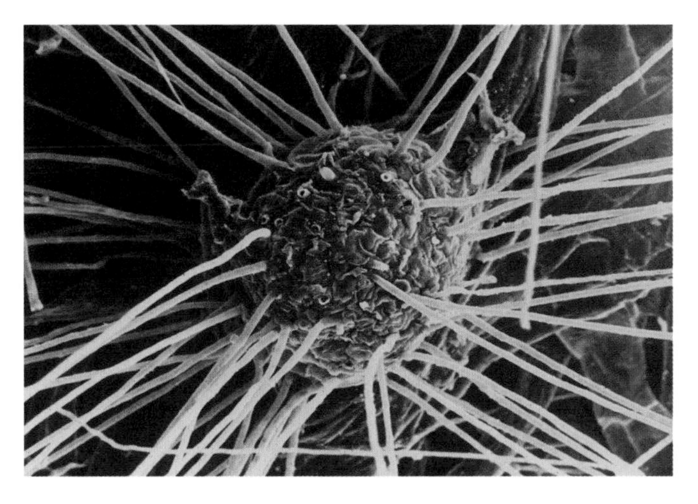

The ectomycorrhizal fungal hyphae on this pine root move out into the soil and explore for and mine nutrients to bring back to the plant in exchange for carbon.

times more water is available to plants when their roots associate with mycorrhizal fungi and form mycorrhizae.

Mycorrhizal fungi create water storage structures in the roots where they are needed. The net of mycorrhizal fungal hyphae branches surrounds plant roots and holds water, acting as a storage reservoir. The interface between some hyphae and the host plant root cells also holds water, as do vesicles, those special storage organelles inside the roots. This is a significant benefit of mycorrhizae. It is also important from an ecological standpoint, as agriculture accounts for about 70 percent of all water use worldwide, and mycorrhizal plants require less water during dry periods.

Protection from pathogens

Mycorrhizal fungi create physical barriers around, and even within, roots, which can prevent some pathogenic organisms from successfully infiltrating the host plant. Mantles, layers of ectomycorrhizal hyphae surrounding some roots, can also make roots difficult to penetrate. If a pathogen does penetrate a root, it is often stopped by the internal structure of mycorrhizae.

Pathogens also have a much more difficult time foraging when they are competing for nutrients with mycorrhizal fungi. These fungi are intense competitors for nutrients in the soil and thus limit access to vital resources needed by pathogens. In addition, as fungal hyphae digest and ingest, they can produce metabolites, chemicals that decay and act as deterrents to other organisms' foraging, and some can even produce their predator's specific defensive and offensive chemicals.

Improved soil structure and carbon storage

As mycorrhizal hyphae weave through the soil, they bind particles together and create soil structure. Mycorrhizal fungi produce sticky exudates that help unite individual soil grains into water-stable aggregates surrounded by pore spaces, which are essential for the movement of air and water. In addition, the fungi produce glomalin, a carbon-laden compound that remains in the soil even after the fungi die. Glomalin molecules are a major source of carbon in the soil.

One of the most amazing things about mycorrhizal fungi is their ability to associate with more than one host plant at the same time—in other words, their networks can be shared among plants, even plants of different species. As a result of this feat, mycorrhizae can benefit entire forests, as the larger trees literally feed and protect the smaller trees through an interconnected mycelial network. And when one plant dies, many of its nutrients are returned to the network and flow toward other plants.

Protectionism

Some mycorrhizal fungi can make it difficult or impossible for plants other than their hosts to grow, which slows down or stops successional changes. The fungi use several mechanisms for this—some produce chemicals that weaken or counteract the chemical components

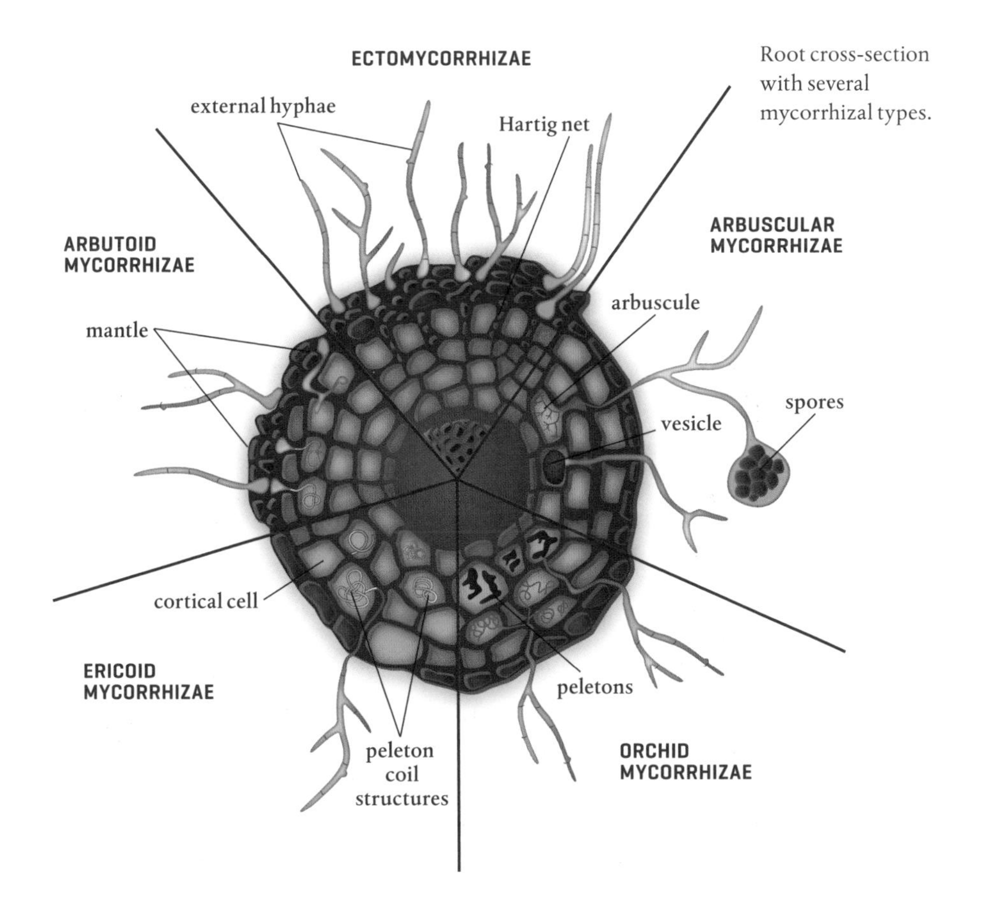

Root cross-section with several mycorrhizal types.

necessary for the invaders' survival, while others influence the microbes in their vicinity with metabolites that attract or repel other plants. They also compete for or provide food sources to protect their host plants.

MYCORRHIZAL TYPES

Mycorrhizal fungi are commonly divided into two groups according to how the fungal cells associate with plant cells. The hyphae of endomycorrhizal fungi penetrate the cell wall, but they do not enter the cell beyond the plasma membrane; these mycorrhizal types are most often associated with the roots of vegetables, grasses, flowers, shrubs, and fruit and ornamental trees. The hyphae of ectomycorrhizal fungi do not penetrate all the way through the cell wall; they form ectomycorrhizae mainly with conifers and some deciduous trees such as oaks.

Endomycorrhizal fungi fall into three main subgroups, each of which has adapted to particular types of host plants, after which the relationship is named: arbuscular, ericoid, and orchid mycorrhizae. Of these, arbuscular mycorrhizae are by far the most common and widespread type. They are of great importance to growers.

ARBUSCULAR MYCORRHIZAE

About 70 percent of all terrestrial plants form the same arbuscular mycorrhizal associations found in fossils from the Devonian period,

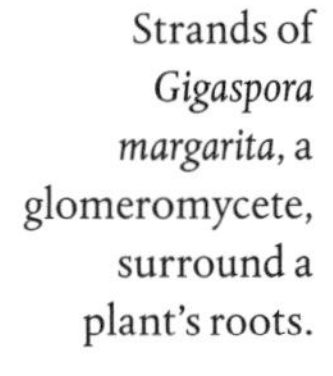

Strands of *Gigaspora margarita*, a glomeromycete, surround a plant's roots.

about 420 million years ago. All arbuscular mycorrhizal fungi are members of the phylum Glomeromycota and form the dominant type of mycorrhizae. Although there are only about 230 species of glomeromycetes, they form mycorrhizae with more than 400,000 different plants. These fungi are not very host-specific and form other associations in nature, including those with many liverworts and mosses. They are not discriminating.

Although an arbuscular mycorrhizal fungus penetrates the plant root's cell wall, it never penetrates the plasmalemma. Instead, the cell forms a membrane that surrounds the fungal hypha, enclosing it in an envelope of sorts. The plant continually builds this membrane as the fungus grows, branches, and produces arbuscules, the tiny, finely branched clusters of hyphae. The enveloping structure creates a cavity into which the two partners deposit their molecular payloads. Many arbuscular mycorrhizal fungi form temporary intercellular vesicles that store water and nutrients and sometimes produce spores.

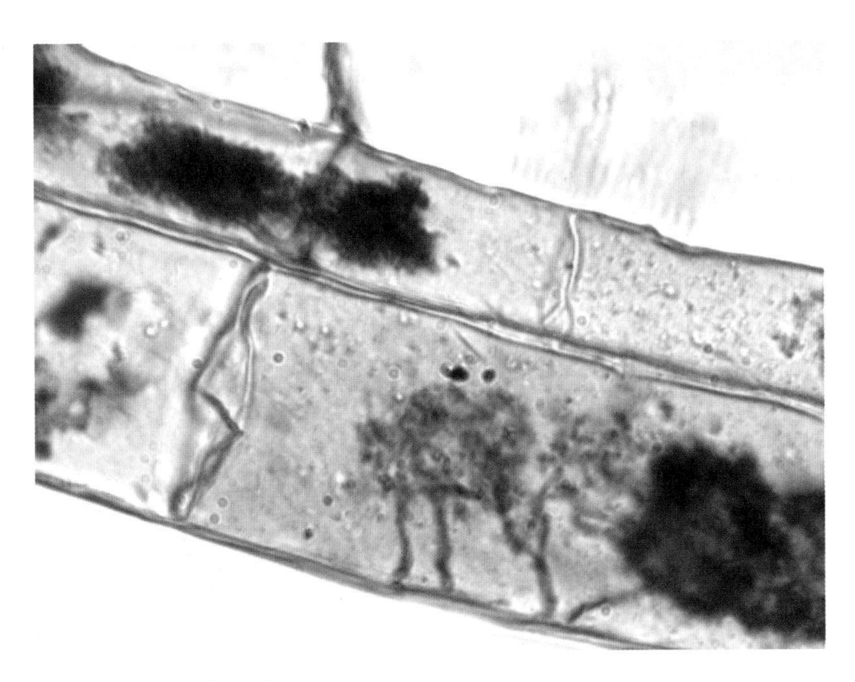

The microscopic arbuscular system;
the dark smudges are arbuscules.

How arbuscular mycorrhizae form

The process of forming arbuscular mycorrhizae is initiated by the plant when soil nutrient conditions, specifically phosphorus levels, are low. Such conditions increase the production and release of strigolactones from the host plant's roots. These specialized hormones attract fungal spores or hyphae in the soil. Once the strigolactones are discovered by the fungus, they guide the fungal hyphae to the host plant's roots. An arbuscular mycorrhizal spore has about seven to ten days to reach a root to get carbon before its own on-board supply runs out and it dies.

Strigolactone molecules are sesquiterpenes, a subset of the terpene group of chemicals, which are found in plants and insects and provide defense against plant–insect and plant–fungus interactions. (Sesquiterpene molecules are also present in the essential oils of products such as myrrh, sandalwood, and cedarwood.) These hormones are released in very low concentrations, which indicates a high susceptibility on the part of the fungus for detecting the signal.

Once discovered by germinating arbuscular fungi, strigolactones cause fungal hyphae to undergo extensive branching. This increases the number of hyphal tips and thus improves the chances of timely contact with the plant roots. Strigolactones also guide the fungi to grow toward

An image from a scanning electron microscope shows an individual arbuscule, which resembles a tiny tree.

RELATED SIGNALS

The signaling molecules produced by mycorrhizal fungi are in the same chemical family as the signaling molecules produced by rhizobia bacteria during the formation of nitrogen-fixing nodules on plant roots. The process of forming nodules involves Nod genes, which are also involved in detecting and reacting to Myc factors. Ponder, for a moment, that these two very different processes, nodulation and the formation of mycorrhizae, use similar chemicals to induce similar plant responses in perhaps common genes.

The hypha of an arbuscular mycorrhiza.

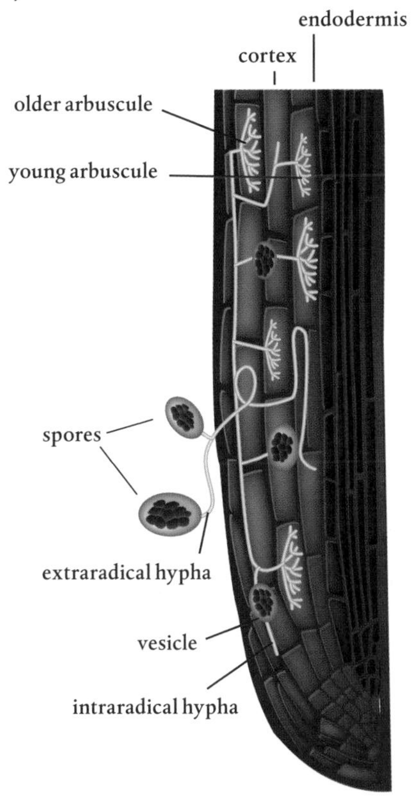

the roots and aid them in the formation of an appressorium, the penetrating mechanism used by fungi to enter the roots. But this is not a one-way dance. The fungi also produce and send chemical signals necessary to prevent the plant from turning its defenses against the incoming mycorrhizal fungi.

These fungal-produced signals, or Myc factors, have been identified as lipo-chitooligosaccharides, a big word that represents an important chemical group that facilitates the close interaction of rhizobia bacteria and leguminous plants. Even low concentrations of lipo-chitooligosaccharides can be detected by plants. When they are detected by plant roots, root growth is stimulated. In fact, this stimulation is

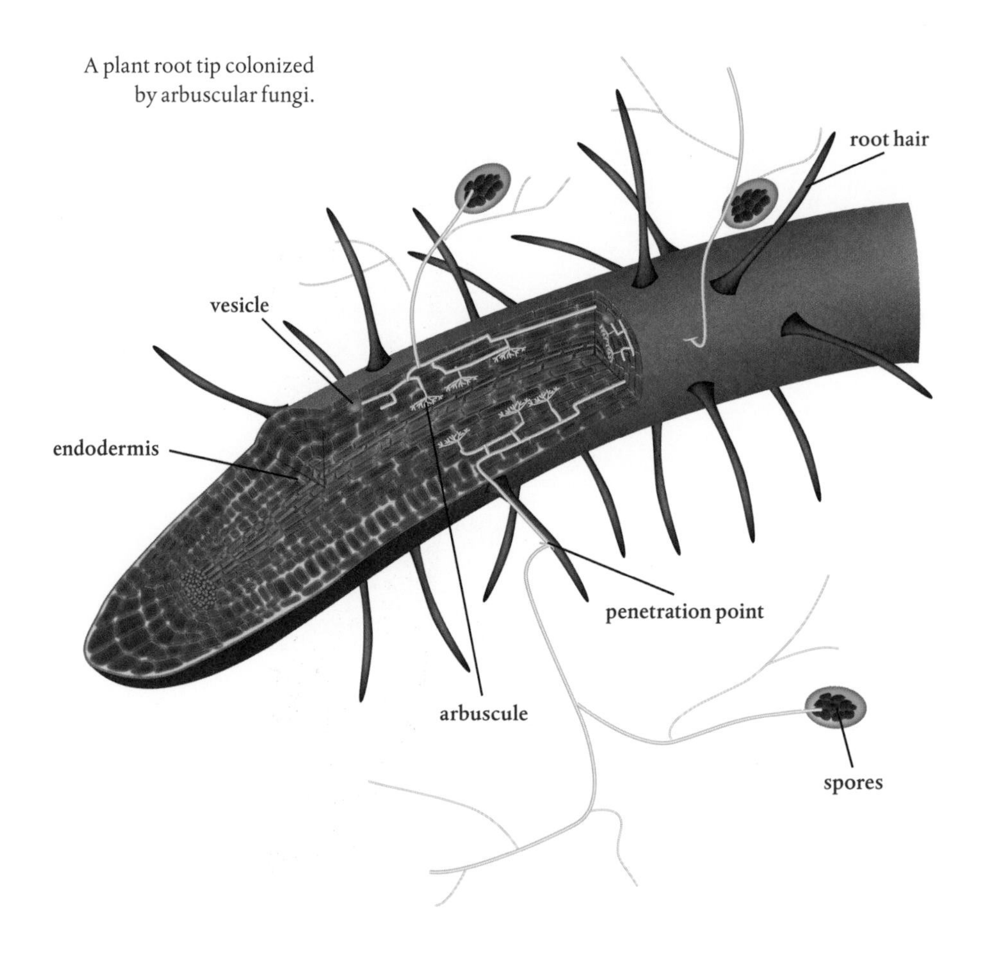

A plant root tip colonized by arbuscular fungi.

important enough to encourage scientific research targeting the development of synthetic molecules that will do the same thing.

When one of the hyphal tips comes into contact with a root as a result of the signaling, an appressorium forms and the epidermis in the root is penetrated. The appressorium is a specialized cell that produces a peg, which is pressed into and then grows into the root. What follows is a coordinated dance between the fungal hyphal membrane and the plant's cell membrane. As the arbuscule develops, its cell wall thins and loosens. In the meantime, vacuoles in the plant cell break up, and the resulting pieces of double membrane concentrate around the invading fungi. These membrane bits merge with the existing plant cell plasmalemma. The hyphal tip is completely surrounded, so it never penetrates the membrane. What forms is an apoplastic cavity that is not directly connected to the inside of the plant cell, where nutrients can be transferred in both directions.

At 10 micrometers wide, a hypha can end up supporting an arbuscule with branches less than 1 micrometer in diameter. These structures are short-lived, and after a few days they start to fall apart. The hypha remains, however, sometimes for months or even years, as do the vesicles formed with the arbuscular features. Vesicles start to form at the same time as the arbuscules, between and inside the host cell walls. They can remain in the root long after the arbuscules are gone.

Meanwhile, as the fungi grow into the soil, they create extraradical hyphae, the fungal network outside the root that increases the surface area for absorption. This mycelial network can enable a plant root to access 100,000 times more soil than a root can access without a mycorrhizal association.

Arbuscular fungal hyphae are almost always nonseptate. If there are septa, they are few and far between. Nuclei and cytoplasm flow throughout the fungal network, which under a microscope can look at lot like the Los Angeles freeway system at rush hour.

Arbuscular mycorrhizal spores

Arbuscular mycorrhizal fungi produce very distinctive spores on the roots of the plant and in the soil. For most fungi, a spore 5 micrometers in diameter is big, but arbuscular mycorrhizal spores are gigantic by comparison: they can measure up to 1000 micrometers in diameter,

THE ROLE OF NUTRIENT TRANSPORTERS

For a while, scientists believed that the arbuscules in host roots disappeared after a couple of weeks, and as they were absorbed by the host, nutrient transfer from the fungus occurred. In the mid-1970s, however, scientists learned that nutrients are transferred out of the fungus into the apoplastic interface created between the fungi and plant and are then transported, via high-affinity transporter proteins, through the plant host's cellular membrane and to the plant's vascular system, where they are needed. The plant membrane is in full control of what enters the plant.

The interfacial apoplast, the space made when the plant's plasmalemma envelops the invading hyphae, is more acidic than the cytoplasm in both the fungus and the plant cells that surround it. This is because it is full of hydrogen ions (H+), the measure of which results in the pH level. In addition, some high-affinity transporter proteins attract and transport phosphorus in the form of phosphate molecules. These ions are further evidence of the presence and operation of transporters used by both plant and fungi that proves nutrient transfer is occurring.

Much of the fungus's energy goes toward making embedded proteins so that it can eat and so that it can transfer some of what it takes in to its host. The host, of course, has a similar system of membranes that allows carbon to be transferred to the fungi. Because scientists can study cells on a molecular level, they have identified the specific transport proteins embedded in the two organisms' membranes. In addition to phosphate transporters are carriers for ammonium (NH_4+), nitrate (NO_3–), and H+ -ATPases (which help in synthesizing ATP). These carriers have been identified in mycorrhizal host plant cell membranes as helping fungi in the mycorrhizal nutrient transfer process, along with chitinases, hydroxyproline-rich glycoproteins, and aquaporins.

Once inside the fungi, plant nutrient molecules are transported to vacuoles and moved to arbuscules. These are complex transactions. Phosphate molecules, for example, go into vacuoles, where they are linked into long chains.

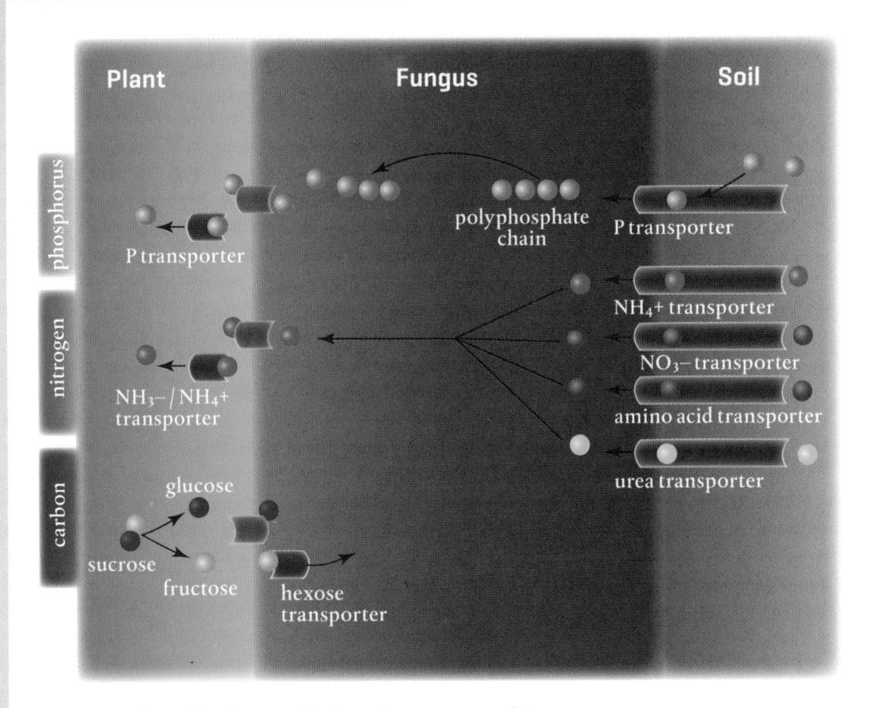

Fungi digest food externally by releasing powerful enzymes that break down food into nutrients that are absorbed in much the same way as plant roots take up nutrients.

These polyphosphate chains are then dumped out of the vacuole and move into the host's cytoplasm. Some of the phosphorus is also used by the fungus for its own biological functions. Urea and various forms of nitrogen, ranging from ammonium to nitrate, are available to plants as well. In exchange, the plant provides the fungus with carbohydrates and sugars including sucrose, fructose, glucose, and hexose.

though most are between 50 and 250 micrometers. You can find many images of arbuscular fungal spores on the Internet to help you to distinguish them by their colors and shapes—they are quite beautiful.

The spores' thick and hard outer shells result in longer shelf life, so to speak, and the ability to withstand abuse, such as passing thorough the digestive track of an insect or a rodent. Because they are so large, arbuscular fungal spores don't travel far in the wind, so catching a ride inside an animal (and being passed through the digestive tract) helps disperse them greater distances.

ERICOID MYCORRHIZAE

Ericoid mycorrhizal fungi form mutualistic symbiotic relationships with members of the plant family Ericaceae, which includes rhododendrons and azaleas (*Rhododendron* spp.) and blueberries and cranberries (*Vaccinium* spp.). These plants grow in acidic, peat-rich soils and make up about 5 percent of terrestrial plant species. The largest fungal group that enters into these relationships are the ascomycetes, which have adapted to the fine root system of ericaceous plants. These roots have an epidermis and a one- or two-cell–layered cortex. The ericoid mycorrhizal fungal hypha penetrates the cortical cell wall and forms a dense coil structure, or peleton.

A veil of hyphae grows over the surface of the roots, but it is finer than

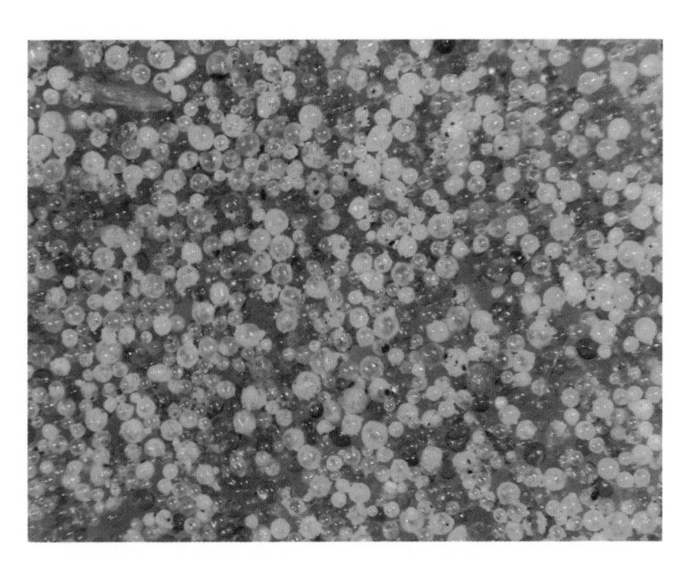

The distinctive spores of glomeromycetes come in many colors and textures.

and not as thick as an arbuscular mantle. Ericoid mycorrhizal fungi do not penetrate as deeply into the soil as other mycorrhizal fungi. They specialize in obtaining nitrogen from organic matter in environments where nitrogen is a limiting factor for plants, producing strong acids that break it up. This has two implications: the fungi free up a lot of hydrogen ions, and the soil environment becomes acidic. Fortunately, ericoid fungi help the plants with which they associate survive in acidic soils that would not be suitable for other plants.

ORCHID MYCORRHIZAE

About 10 percent of the Earth's plant species are orchids (family Orchidaceae), and most of them depend on specialized endomycorrhizal fungi from the Basidiomycota at some point in their lives. Orchid seeds are tiny and do not contain sufficient nutrients to support the growing embryonic plant; they get what they need from the mycorrhizal association.

The orchid seed germinates and sends out a few hairs that are immediately colonized by fungi. The hyphae grow into the root cell epidermis and the cortex, which is thicker than in ericaceous plants, allowing for a complex network of peletons. The peletons live for only a few days and are then absorbed by the plant. These associations can be temporary, disappearing later in the plant host's life or being replaced by another

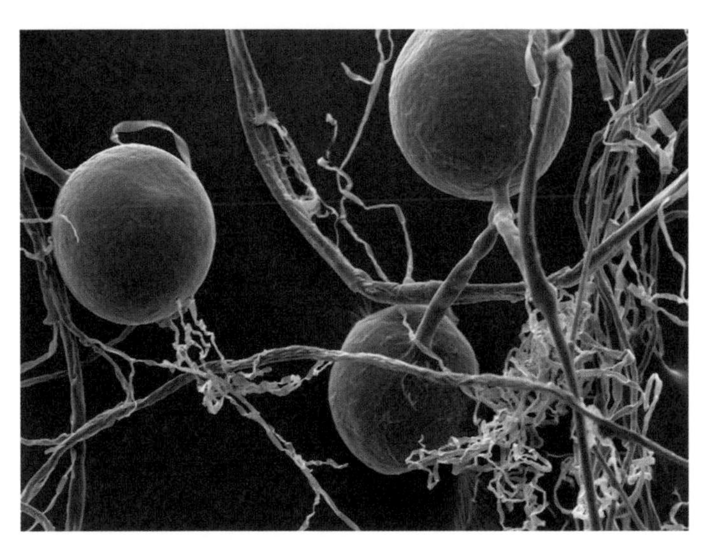

Although some are visible to the naked eye, most arbuscular fungal spores are visible only under magnification.

kind of symbiotic fungi. Some orchids are heterotrophic and do not contain chlorophyll, so they rely on their fungal partners to obtain nutrients throughout their lives.

ARBUTOID MYCORRHIZAE

The fungi that form arbutoid mycorrhizal relationships are the Basidiomycota. The most important host plants in these mycorrhizal relationships are the Pacific Northwest madrone tree (*Arbutus menziesii*), from which the arbutoid name is derived, and manzanitas and bearberries (*Arctostaphylos* spp.). In arbutoid mycorrhizal associations, a mantle surrounds the roots of the host plant, and intercellular Hartig nets are formed. Sometimes, however, arbutoid mycorrhizal fungi will penetrate the root's cell wall, as in endomycorrhizae. The hyphae penetrate the outer cortical cells of the plant roots, forming tiny coils that allow for the transfer of nutrients.

MONOTROPOID MYCORRHIZAE

Monotropoid mycorrhizal fungi were once thought to be part of the arbutoid group, but they do not penetrate the plant cell walls. They colonize plants of the family Monotropaceae, which includes Indian pipes (*Monotropa* spp.), a woodland plant that lacks chlorophyll. In fact, all plant hosts in this group lack chlorophyll. The fungi form an ectomycorrhizal association with trees such as beech (*Fagus* spp.), oak (*Quercus* spp.), and cedar (*Cedrus* spp.) and then form a monotropoid association and transfer some of the trees' carbon to the plants.

Monotropoid mycorrhizal fungi form a dense sheath, or mantle, around the root. These fungi also form Hartig nets, which surround but do not penetrate the root cells. Some individual hyphae form fungal pegs, which do penetrate the cortex. These live inside the host plant for a couple of weeks and then die. The host plant absorbs the fungal hyphae after they die to obtain the carbon it needs.

ECTOMYCORRHIZAE

Ectomycorrhizal fungi are more modern than endomycorrhizal fungi. They evolved to associate with plants about 250 million years ago. Although there are several thousand different types of ectomycorrhizal fungi, only about 5 percent of terrestrial plants form ectomycorrhizal associations with ascomycetes or basidiomycetes. In addition, many of these fungi produce mushrooms next to their plant hosts and are not only recognizable but often of economic and culinary value. Ectomycorrhizal fungi are more acidophilic than other mycorrhizal fungi—that is, they prefer acidic soils with a low pH.

The plant hosts of ectomycorrhizae are always woody plants, generally trees, including pine (*Pinus* spp.), beech (*Fagus* spp.), birch (*Betula* spp.), and myrtle (*Myrtus* spp.). Although there are fewer plant hosts than exist for endomycorrhizal fungi, the plants that are colonized by ectomycorrhizal fungi include some 25,000 tree species found across the globe. This may have to do with the diversity of their fungal symbionts, of which there are a few thousand.

Ectomycorrhizae formation begins with fungi–root contact below the apex of young, actively growing roots. The hyphae first grow on the surface of the root. After a day or two, a mantle is formed around the root, followed by penetration between the host cortex cells, branching, and growth, which results in the formation of the Hartig net between the cells. In angiosperms (herbaceous plants, shrubs, and most trees), this is usually a single cell layer and it does not move out of the root's epidermal

A fairy ring of mushrooms around this white spruce may indicate ectomycorrhizae formation below ground.

cells. In gymnosperms (conifers, ginkgo, and cycads) that associate with ectomycorrhizae, the Hartig net can penetrate the cortex.

As the ectomycorrhizal fungus covers the roots, only a small portion that sits about 1 to 3 millimeters from the growing tip is active. As the root grows, this active zone continues to move outward, and older parts

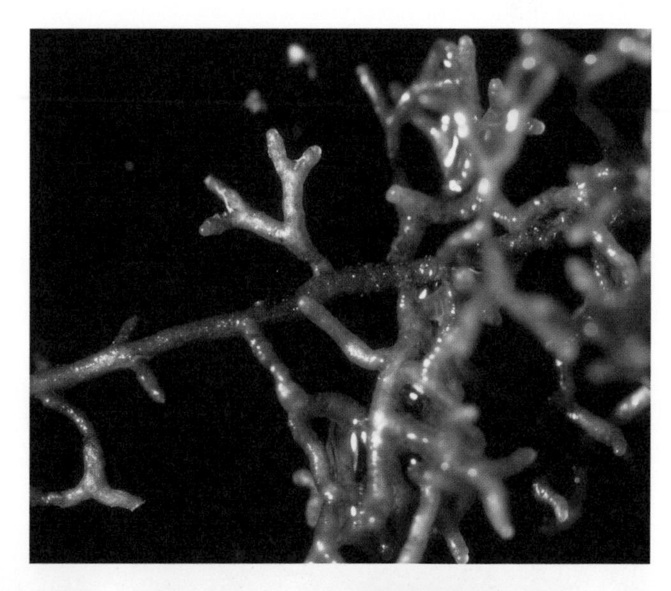

Classic ectomycorrhizae with red pine as the host.

Amanita muscaria (fly agaric) is a fruiting body of an ectomycorrhizal fungus that colonizes birch trees.

of the Hartig net die. Meanwhile, the fungal hypha that spread into the soil from the mantle will grow parallel to one another and branch. They can also merge to form rhizomorphs, hollow structures that store and transport water, or sclerotia, which store nutrients. Rhizomorphs, in particular, can look like roots to the naked eye, and they perform some of the same functions. They can carry nutrients and water, sometimes very long distances.

ECTENDOMYCORRHIZAE

Though rare, some mycorrhizae have the characteristics of both endo- and ectomycorrhizae. This seems to be confined to a few species of deuteromycetes, which are imperfect fungi. Ectendomycorrhizae form in pot-grown nursery seedlings and with some tree seedlings, mostly pines, after forest fires. The host can be either coniferous or deciduous. A thin mantle is formed, and the hyphae of the Hartig net penetrate the cortex cells but not the plasma membrane. Eventually, these associations morph into ectomycorrhizae as the seedling matures.

SEBACINOID MYCORRHIZAE

A special group of basidiomycetes in the order Sebacinales, the sebacinoids are very diverse in terms of their mycorrhizal relationships. Some form endomycorrhizae while others are ectomycorrhizal in nature, or they form ericoid and even orchid mycorrhizae. Sebacinoid fungi can colonize the tree roots of *Eucalyptus marginata*, one of the most common species of native eucalyptus in Australia.

Piriformospora indica, a sebacinoid fungus discovered in desert soils in

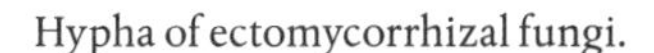

Hypha of ectomycorrhizal fungi.

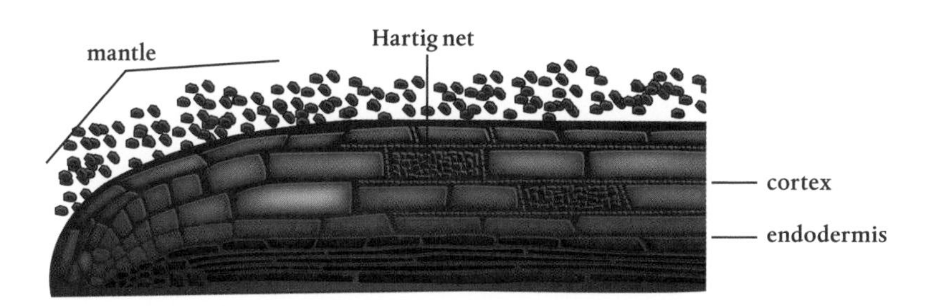

MULTI-MYCORRHIZAE

Rarely, more than one species of fungi can enter into a mycorrhizal association with a single host plant. Most poplars (*Populus* spp.) can form either ectomycorrhizal or arbuscular mycorrhizal relationships, depending on the species, but *P. tremuloides* can be associated with both types simultaneously. This is also the case with other trees in the willow family (Salicaceae), as well as alder (*Alnus* spp.) and eucalyptus. The same is true, though far less often, for ericaceous plants. This phenomenon is not often studied because of its rarity. It is fascinating, however, to think that a host plant might be experimenting with or testing different fungi for future use.

northwest India, could have huge agricultural and commercial value. When scientists grew it in the lab, it successfully colonized several crop plants, including corn, barley, wheat, parsley, and poplar. Experiments have shown that *P. indica*–inoculated plants are more resistant to some root pathogens and diseases.

PLANTS WITHOUT MYCORRHIZAL ASSOCIATIONS

Plants that do not form mycorrhizae are the exceptions. Seed and plants of the families Brassicaceae (such as mustards, kales, and cabbages), Caryophyllaceae (such as pinks and carnations), Chenopodiaceae (such as amaranths and goosefoot), Polygonaceae (such as knotweed and buckwheat), and Portulacaceae (such as purslane) germinate and remain healthy without mycorrhizal associations. These plants usually have extensive and quick-growing specialized root systems that make the most of limited nutrients in poor soils such as sand. Some nonmycorrhizal plants have very long roots covered with fine root hairs, which can explore soil pores better than most other plants, without the help of mycorrhizal fungi. Others have developed carnivorous capabilities and get nutrients from aboveground organisms such as insects. Still others, such as epiphytes (most orchids), can grab nutrients from the air, and some have evolved into parasites.

Proteas (Proteaceae) and lupines (*Lupinus* spp.) have cluster roots, closely spaced lateral roots that are densely covered with root hairs. These roots form near the soil surface so they can feed in the duff layer. To obtain phosphorus without mycorrhizal fungi, they support lots of phosphate-solubilizing bacteria in their rhizospheres. Their exudates are full of organic acids that decay organic matter.

Some members of the Cyperaceae (sedges) have dauciform roots—swollen, carrot-shaped lateral roots that are densely covered in root hairs. Capillaroid plants such as rushes (Restionaceae) form a mat of root clusters that are densely covered in long root hairs at the soil surface. These roots pump out lots of organic acid exudates that help them take up the nutrients they need.

Some sedges, rushes, and other monocots in Western Australia have sand-binding roots, in which sand literally sticks to the roots and covers them completely. Scientists hypothesize that the attached sand somehow enhances nutrient uptake.

GENERAL LOCATIONS OF MYCORRHIZAE TYPES

Determining the succession and location of mycorrhizal fungi and plants in the soil can be a chicken-or-egg situation: the type of mycorrhizal fungi present in the soil affects the distribution and diversity of

The cluster roots of the red pincushion protea, *Leucospermum cordifolium*, are not colonized by mycorrhizal fungi.

the plants that grow there, but at the same time, the plants may control the local distribution of the fungi by associating or failing to associate with a particular fungal species.

If no mycorrhizal fungi are present in the soil, nonmycorrhizal plants will grow there. Eventually, facultative mycorrhizal fungi, which are parasites or saprobes, will move in. Semidependent host plants will then become established and will eventually take over. As decay increases, obligatory mycorrhizal fungi will colonize the area, along with completely dependent host plants.

Studies have shown that, in response to the presence of mycorrhizal fungi, plant diversity usually increases. Studying changes in the type of mycorrhizae in a particular area can tell scientists which way the area is headed. Nevertheless, it is clear that the presence of a particular mycorrhizal fungus influences the ability of particular plants to live in particular soils. Of course, the ability of mycorrhizal fungi to change the environment means that some plants may slow down or cease to grow there. So it goes both ways.

Mycorrhizal fungi can also be grouped based on the geographic location in which they grow, which in turn is based on soil type, the availability of nutrients, and the prevailing climatic conditions. For example, some particular ectomycorrhizal tree hosts grow only in particular areas, and that is where you will find their associated mycorrhizal fungi.

Soils in subarctic regions have extremely low pH levels. Peat dominates there, as do ericoid mycorrhizae. Ascomycete fungal partners penetrate the root's cortex cells, but they don't send much mass into the soil. They do, however, mineralize nutrients near the roots that the host plant could never access alone, producing the inorganic ions needed by the host. Some tundra mycorrhizal fungi provide organic nitrogen directly to their host, a feat that harkens back to the days of the Humus Theory (which held that plants ate organic matter) and bucks the norm of taking up inorganic forms of nitrogen. (If this process could be translated into agricultural plants, it could change the way the world is fed.)

In conifer forests and hardwood forests, with warmer temperatures, higher pH, and more available nutrients, ectomycorrhizal fungi and their associations predominate. These fungal hyphae send out lots of hyphal mass, which not only brings back lots of nutrients to the hosts but also mineralizes a tremendous amount of organic material, making it available to other plants and fungi.

In temperate grasslands with warmer temperatures, where the soil pH is even higher and nitrates are the predominant form of nitrogen, arbuscular mycorrhizae prevail. The huge increase in surface area available to the system as a result of mycorrhizae enables the host plants to meet their nutrient needs.

MYCORRHIZAL BEHAVIOR

Scientists have demonstrated that nutrients are shared throughout mycelial networks, both between different and among the same plant species. In fact, several different mycelial networks can fuse together and act as one. A mother tree can thus support other trees throughout the forest, preferentially distributing more nutrients to the younger ones. Cooperation exists between different plants' mycelial networks. This is amazing when you consider that each plant may also contain substances that can inhibit the growth of another.

It is clear from genome sequencing that mycorrhizal fungi contain special genetic expressions that cause them to function as they do. Some genes affect nutrient uptake, others manage the production of signaling proteins used in controlling the delicate dance between mycorrhizal

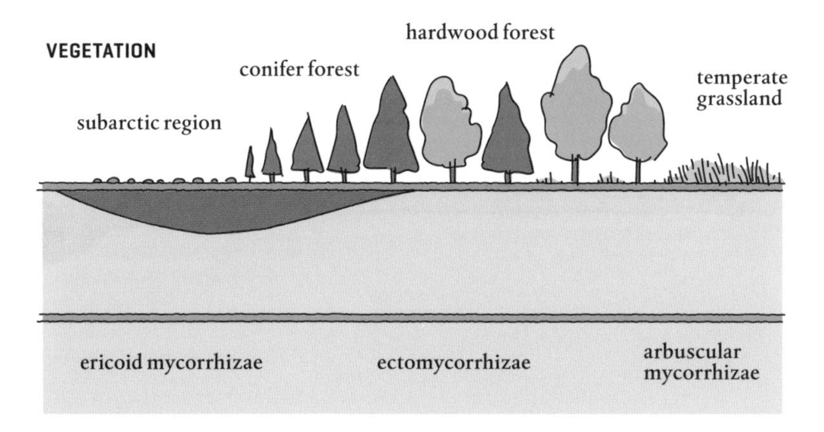

Locations of vegetation and corresponding mycorrhizal types.

BACTERIAL DNA IN MYCORRHIZAL SPORES

When scientists examined the DNA that makes up arbuscular mycorrhizal spores, they discovered that a significant amount of the spore mass comprises symbiotic bacteria. One study indicated that as much as 40 percent of the total spore mass contained DNA from bacteria. This is of interest because the DNA of a fungus can be exchanged with the DNA of a bacterium; this may occur in the spore as a result of environmental conditions.

symbionts, and others help the fungi reach the plant without triggering its defense mechanisms. These subtle genetic differences, however, may result from DNA exchange with other mycorrhizal fungi or with other organisms, such as bacteria, in the mycorrhizosphere or the fungal spore. Perhaps this occurrence depends on the conditions to which the spore is exposed and in which the fungus will need to live.

Knowing how mycelial networks function and figuring out how we can duplicate their operation could be a major advance, especially with regard to food crop production. Scientists have discovered that microorganisms exchange genetic material. They can create maps of the genetics of particular mycorrhizal fungi and study these in relation to a whole range of biological influences. This new science, transcriptomics, enables scientists to identify mycorrhizal strains that carry specific traits that can benefit a specific crop or plant. We still have a lot to learn, and perhaps we will someday use all this knowledge to benefit humankind.

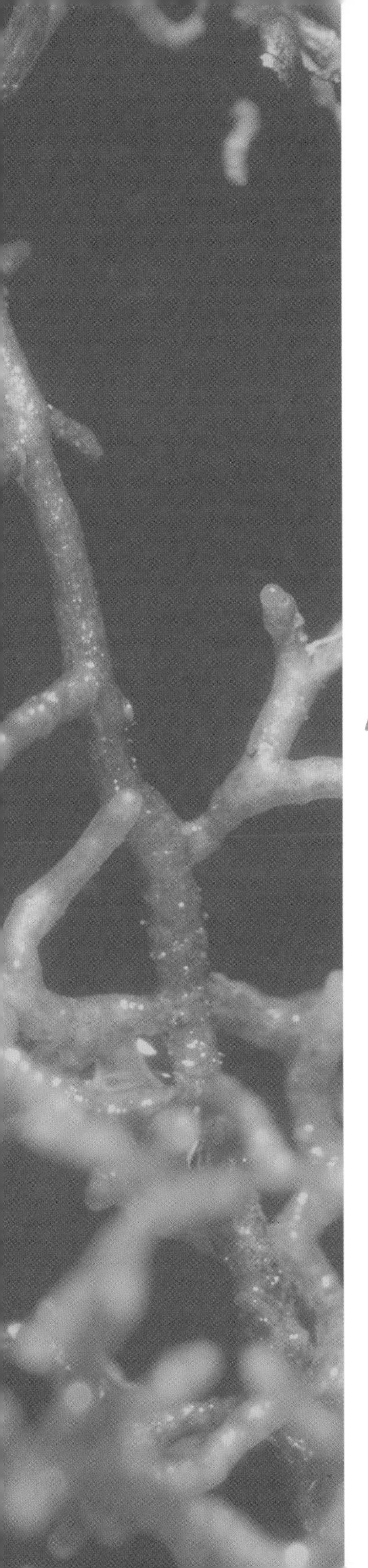

Mycorrhizae in Agriculture

As human populations evolved from hunter-gatherers to farmers, most of the crops they planted depended on mycorrhizal fungi. Early agricultural practices caused little disturbance to soil, and even when farmers began using beasts of burden to break up the soils, the presence of mycorrhizae was sustained. Unfortunately, modern agricultural practices rely on heavy machinery, heavy soil disturbance, periods of fallow, artificial chemical fertilizers, pesticides, herbicides, fungicides, fumigants, and sometimes soil sterilization. These practices, as well as others, disrupt and destroy the establishment of mycorrhizae in agricultural soils.

Commercial mycorrhizal fungi have been available to the green industry since the 1990s and to the home gardener, in small quantities, for almost as long. Successful efforts at producing commercial mycorrhizal fungi have enabled growers to inoculate on a large scale, and this is opening up exciting new possibilities in the realm of agriculture, with many ramifications for the future. Still, growers throughout the world are focused on the bottom line. They need to be convinced that using mycorrhizal fungi—in particular arbuscular mycorrhizal fungi, which

CHEMICAL RECKLESSNESS

In an article for *The Atlantic*, "Healthy Soil Microbes, Healthy People" (11 June 2013), longtime friend and mycorrhizal scientist Mike Amaranthus (with coauthor Bruce Allyn) compares the increasingly disastrous impact of overuse of antibiotics and highly processed food on the human gut microbiota to the impact on the soil's microbiota as a result of using chemicals: "We have recklessly devastated soil microbiota essential to plant health through overuse of certain chemical fertilizers, fungicides, herbicides, pesticides, failure to add sufficient organic matter (upon which they feed), and heavy tillage." He adds that "reintroducing the right bacteria and fungi to facilitate the dark fermentation process in depleted and sterile soils is analogous to eating yogurt (or taking those targeted probiotic 'drugs of the future') to restore the right microbiota deep in your digestive tract."

most often partners with crop roots—is more efficient and profitable than using chemicals. It all starts with learning what mycorrhizae can do in an agricultural setting.

For many years, scientists believed that mycorrhizal fungi were ubiquitous, and that the spores of all fungi traveled around the world in wind

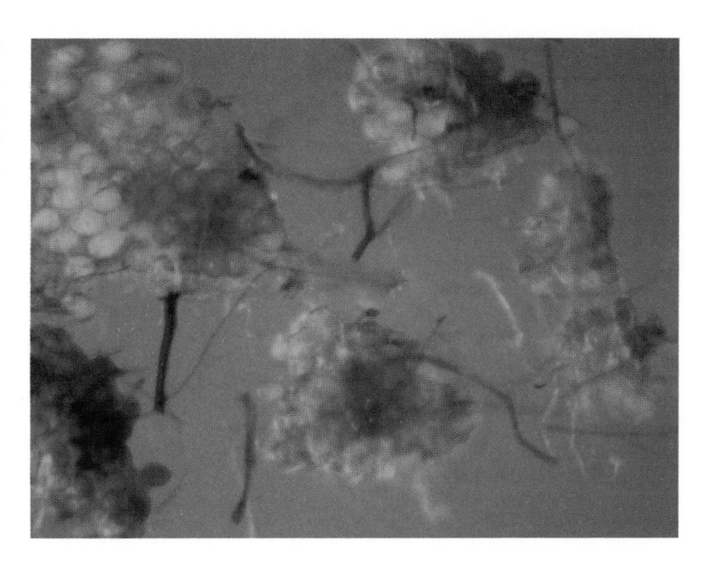

Some arbuscular mycorrhizal fungal spores resemble clusters of salmon eggs.

and air currents, so they didn't need to be added to soil. Now we know that arbuscular mycorrhizal spores are large and relatively heavy, unlike the spores of ectomycorrhizal fungi. They don't travel that far. We also recognize that, under the right circumstances, mycorrhizal fungi need to be added to agricultural soils and managed to benefit most crops. Study after study has demonstrated that plants inoculated with mycorrhizal fungi show superior growth and health. In the future, the use of arbuscular mycorrhizal fungi will become as widespread as the use of fertilizers today.

NUTRIENT UPTAKE ENHANCEMENT

Within the soil, a plant's roots are limited to a small area for absorption of nutrients; arbuscular mycorrhizal fungi grow out beyond the depletion zone, where the roots have already removed the available nutrients and water. Fungal hyphae can grow long distances and can include rhizomorphs that extend much farther than the root and rapidly return water to the plant. The increased surface area provided by the fungi also improves the roots' nutrient absorption.

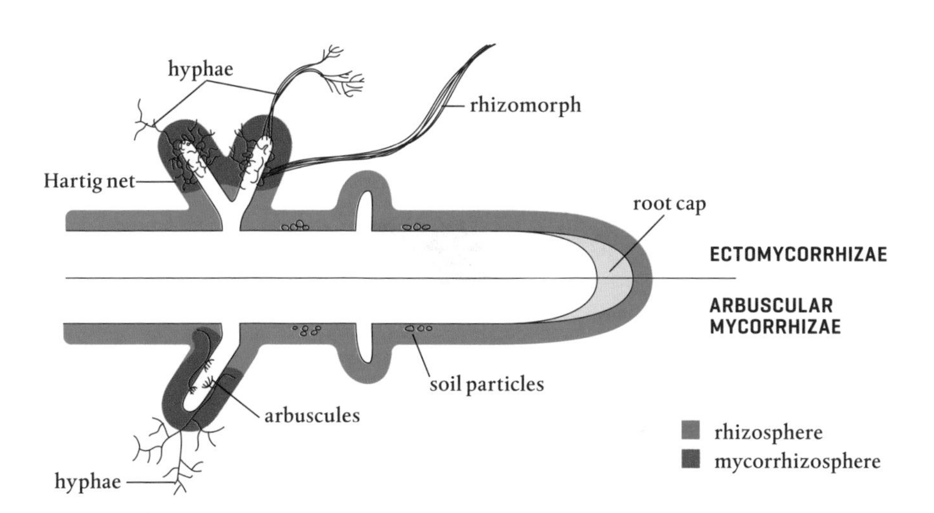

Hyphae extend from the root's rhizosphere into the mycorrhizosphere, where they find new sources of nutrients beyond the depletion zone.

PHOSPHORUS AND THE ENVIRONMENT

Phosphorus is a nonrenewable resource—and known reserves are expected to be depleted within 50 to 100 years. Many people are concerned about a looming crisis as the need to produce more food is complicated by the decreasing availability of this valuable mineral. In addition, widespread application of phosphorus in agricultural soils has created global environmental consequences, as significant amounts of phosphorus are washed or eroded away and wind up in streams, where they kill aquatic life and destroy the ecosystem. For these reasons, finding a substitute for the application of phosphorus fertilizers is an important quest. Mycorrhizal fungi may provide the best solution.

Microorganisms in the soil food web immobilize inorganic phosphorus by converting it to organic molecules that plants do not take up. During fungal digestion, the release of acids and enzymes break down organic and inorganic matter. Roots alone cannot produce these enzymes, and these nutrients would not otherwise be available to the plant unless some other soil food organism made them so.

Supplies of nutrients are constantly being cycled in the fungal cytoplasm and are thus available to the fungus as well as the host plant when they need it. These nutrients are also stored in fungal cell vesicles and vacuoles. Fortunately, there is enough dilution by cellular activity in both the plant and fungi so that movement of nutrients into the fungal hyphae is not disrupted. The mycelial network created by arbuscular mycorrhizal fungi also represents an ideal storage structure, where areas of lesser concentration of nutrients can serve as a sink toward which nutrients will flow from the soil.

A major benefit of the association between a host plant and mycorrhizal fungi is the plant's increased ability to take up phosphorus, a nutrient that is important because of its presence in ATP, the principal molecule for storing and transferring energy in living cells. Of all the nutrients a plant needs, phosphorus is perhaps the most difficult for a plant to obtain in sufficient quantities without mycorrhizal assistance.

In fact, studies have revealed that 75 to 95 percent of the phosphorus fertilizer applied to crops is not taken up by the plants. Phosphorus molecules are chemically bound to soil because negatively charged phosphate anions are attracted to and held by positively charged cations in the soil—in this case, iron, aluminum, and calcium cations. Phosphorus ions are quickly absorbed and held by clay soils, which are loaded with cations. Arbuscular mycorrhizal fungi provide their plant partners with up to four times more phosphorus than the plant is able to obtain absent the relationship. This is an impressive feat, for sure.

By 1957, studies suggested that mycorrhizal associations resulted in greater phosphorus uptake. Many scientists of the time were testing mycorrhizal fungi in soils that contained lots of phosphorus, however, believing that the fungi needed to pool all the nutrients before distributing them to the host plant. These fungi did not fare well, were slow to colonize, and didn't show the expected benefits. A controversy developed in the scientific world as to the efficacy of mycorrhizal fungi in growing crops. Then, in the mid-1960s, scientists realized that mycorrhizal fungi would bring phosphorus to plants only if the supply in the soil was low and the plant alone could not reach it. Too much phosphorus had detrimental effects on mycorrhizae formation.

In the 1990s, scientists started using special mesh dividers to isolate the root systems of plants, enabling controlled experiments on mycorrhizae to be readily observed. They learned that the fungi actively take up phosphorus (and nitrogen, copper, nickel, iron, magnesium, manganese, zinc, and water) to feed the host plant.

The fungi may not be solely responsible for the increase in phosphorus uptake, however. Although the fungi produce the necessary enzymes to free phosphorus from chemical bonds in the soil, all manner of biological helpers associate with mycorrhizal fungi in the mycorrhizosphere. Soil organisms attracted by the exudates from the fungi or the remnants of its meals include phosphate-solubilizing bacteria that help break some of these chemical bonds, making it easier for the fungi to take up the phosphorus. The same process, with different organisms, may impact the intake of other nutrients.

Arbuscular mycorrhizae can also enhance the uptake of phosphorus, nitrogen, sulfur, copper, zinc, boron, iron, magnesium, and manganese for their hosts. These nutrient metals are relatively immobile in the soil and must attach to anions in clay and organic matter before they can be absorbed by roots alone. Mycorrhizal fungi venture out and get them, sometimes releasing them from their substrates and always increasing the absorption of these ions, sometimes up to 60 times. Considering the cost of fertilizers, fungi are clearly a viable alternative.

Nitrogen taken up by arbuscular fungi is available to plants in several forms, usually as ammonium (NH_4+) but also as nitrates (NO_3-) or in several amino acids. Ammonium, like many other cations, is not very mobile in soil and can be depleted around the root zone. Mycorrhizal fungi spread through the soil to access new supplies. Once nitrogen is inside the fungal membrane, it is probably assimilated into amino acids. Then the nitrogen-containing molecules are circulated in the cytosol (the liquid part of the cytoplasm) for use by the fungus. Some is converted back to ammonium and transported into the apoplastic interface, the area outside a plant root cell's plasma membrane where diffusion into the plant cell occurs. Some nitrogen is also transported across membranes as proteins.

Another important nitrogen impact results from arbuscular mycorrhizae: the establishment of mycorrhizae increases the activity of nitrogen-fixing nodules in leguminous plants. Phosphorus is needed in the nitrogen fixation process, and its increased presence, thanks

Both the number of nitrogen-fixing nodules and the amount of nitrogen they produce are increased by the presence of arbuscular mycorrhizal fungi (the hazelnut is for scale).

WHAT DO THE FUNGI GET?

We know all the benefits a host plant receives from partnering with a mycorrhizal fungus, which helps the plant meet up to 80 percent of its phosphorus and nitrogen needs. In exchange, the plant provides carbon to its fungal partner, without which it would die.

Plants manufacture carbon through photosynthesis and convert it to sucrose, which is transported to other parts of the plant. With an arbuscular fungal association, however, the carbon provided from the plant is in the form of hexose: the plant produces the necessary enzymes to convert sucrose into hexose for the fungi. Once the hexose moves into the apoplastic interface, it is actively transported through the fungal membrane, and more energy (ATP) is expended to move it through a plasma membrane protein into the fungal cytoplasm. Inside the hypha, the hexose is converted to lipids, insoluble fatty acids that are very different from sugars. These are then cycled and translocated into the extraradical network outside of the root.

Is this a fair trade? It may not appear that the mycorrhizal fungus is sacrificing much, considering what it gets in return. From the plant's perspective, this is an energy-driven relationship and there are sacrifices. Up to 20 percent of a host plant's manufactured carbon can be transferred to its fungal partner, which uses the plant's energy to increase its activities to pay back the plant. It has to keep the host plant alive, after all. The sacrifices must be worth the effort, because the plant–mycorrhizal fungi relationship has continued for more than 400 million years.

to mycorrhizal fungi, increases the amount of bacterial colonization in the root nodules of legumes. In fact, when arbuscular mycorrhizal fungi colonize legume roots, the very number of nitrogen-fixing nodules increases.

PATHOGEN RESISTANCE

Mycorrhizal colonization can help plants resist infection from parasitic nematodes, fungal pathogens, and other disease-causing organisms. Because colonized plants are strong and healthy, they can resist

THE MYCELIAL NETWORK: EARTH'S NATURAL INTERNET

We know that mycorrhizal fungi and their associated host plants both benefit by exchanging nutrients and other compounds. There may be more to the story, however, that suggests an even deeper connection between plants and the mycelial network. Plants have evolved defense adaptations that help improve their survival and reproduction, including the release of organic compounds that act as repellents or toxins to insects or herbivores or that reduce the plant's digestibility. Studies have shown that mycorrhizal plants may also send chemical signals throughout the extraradical mycelial network to communicate with neighboring plants and warn them of pest attacks.

Experiments show that the communication signaling over the mycelial network results in the formation of defense-related genes and activation of other defense compounds. In addition, studies show that mycelial networks transmit not only signals that activate defensive mechanisms in neighboring plants but also toxins that prevent neighboring plants from thriving and thus competing for valuable nutrients.

The ability of different mycelial networks, with different mycorrhizae, to communicate is an astonishing feat. Noted mycologist and author Paul Stamets calls mycelia "Earth's natural Internet."

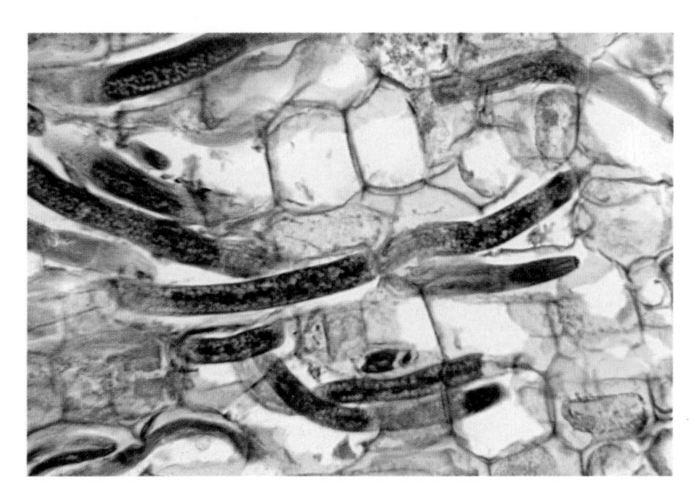

Studies have shown that mycorrhizal root colonization can help reduce the damage caused by root nematodes, including the common burrowing nemotodes shown here.

or tolerate root rot pathogens such as *Fusarium*, *Rhizoctonia*, *Pythium*, and *Phytophthora* as well as stem diseases such as *Verticillium*. In addition, the fungi compete for the limited amounts of nutrients in the soil, leaving few nutrients available to support pathogenic organisms. Colonization by mycorrhizal fungi causes plant roots to branch more and to become shorter and thicker, and because this robust fungi–root system is more difficult to penetrate, the effects of some root rot problems are lessened. As the cell walls of mycorrhizal roots thicken and strengthen, they become more difficult to penetrate. When mycorrhizal plants are damaged, they can heal faster by forming wound barriers and clogging stomata, the tiny pores in plant leaves through which gases and water vapor pass. Finally, mycorrhizal plants have stronger vascular systems.

Much study on arbuscular mycorrhizae has involved their interaction with parasitic nematodes. Hundreds of studies have paired specific crop plants with specific fungi to study the impacts of mycorrhizae on nematodes. As a result, we know that arbuscular mycorrhizae provide a robust defense alliance by producing chemicals that reduce nematode reproduction, feeding, and attraction to plant roots. In addition, the mantle that surrounds the root can provide a physical barrier to nematode predators.

Arbuscular mycorrhizal fungi create metabolites that include biocides, antibiotics, and pest-specific chemicals that can attack and destroy harmful organisms. Studies confirm that soybeans, tomatoes,

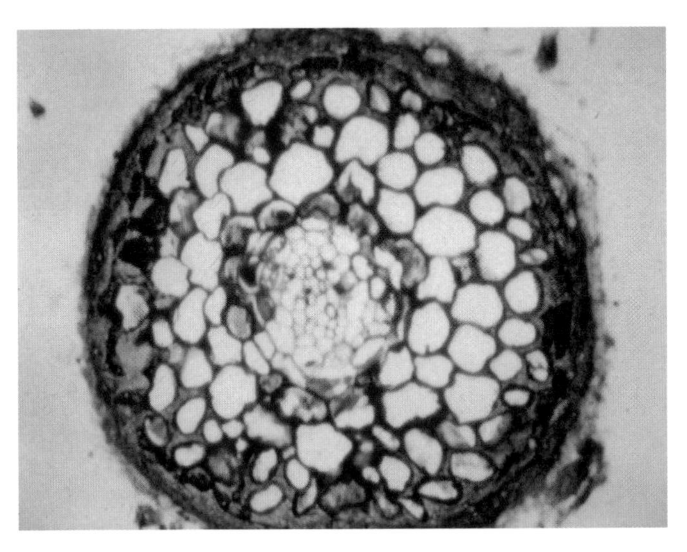

The mantle (in the outer blue area) acts as a physical barrier to nematode predators.

oats, onions, and cotton, for example, are less susceptible to nematodes when arbuscular mycorrhizal fungi are present and mycorrhizae form in the soil.

HORMONE PRODUCTION

The formation of mycorrhizae can increase the production of a variety of plant hormones. These hormones are important signaling molecules that regulate cellular processes, including the formation of flowers, stems, and leaves and the development and ripening of fruit. They also are required before a fungal spore will grow into a mycorrhizal network.

Strigolactones, for example, stimulate the branching and growth of arbuscular mycorrhizae and help increase the probability that the growing fungus will find the host plant's roots. In addition, production of several other hormones is increased: cytokinins that promote cell division and growth in roots, abscisic acid that promotes leaf detachment and induces seed and bud dormancy, ethylene that causes fruit to ripen, jasmonic and salicylic acids that help plants defend themselves against stresses such as pathogens and insects, and growth auxins that orchestrate almost every aspect of plant growth and development. Scientists continue to study the relationship between establishment of arbuscular mycorrhizae and plant hormones because of the potential agricultural applications.

SOIL RESTORATION

Establishing arbuscular mycorrhizae in agricultural soils can reverse the damage caused by modern crop production. Glomalin, the sticky, carbon-based glycoprotein (part sugar, part protein) secreted by arbuscular mycorrhizal fungi, helps strengthen hyphal walls and seal gaps, but its most important attribute, where agriculture is concerned, is the amount of carbon it contains. Glomalin contributes almost a third of the soil's carbon where mycorrhizal fungi are present. The carbon in glomalin molecules far surpasses the amount present even in humic acids, which weigh up to 20 times more. Glomalin's stickiness helps bind soil particles together, adding structure. As a result of the presence of arbuscular mycorrhizae and glomalin, soil aggregation, aeration, and drainage are greatly improved.

Arbuscular mycorrhizae also add organic matter to the soil as the fungi die and decay. These fungi can make up as much as 50 percent of the microbial mass in a given volume of soil. While the fungi are alive, they provide habitat for all manner of organisms that are attracted to their exudates or the protection they afford. These organisms tunnel and burrow through the soil, contributing to soil structure and aeration. When they die, the end result is a stable, carbon-rich agricultural soil that is more resistant to wind and water erosion.

DROUGHT RESISTANCE AND TEMPERATURE TOLERANCE

In times of drought, plants with arbuscular mycorrhizal associations show less stress than plants with few or no mycorrhizae. This could result from several factors: increased soil–root contact, a colonized plant's ability to explore for and find water and greater ability to absorb it, changes in osmotic forces, increased number of aquaporins, or changes in hormonal signaling in mycorrhizae. Or perhaps plants with mycorrhizae are simply bigger and healthier, so they are better able to thrive even in drought situations. Whatever the reasons, experiments have shown the impact of mycorrhizal associations to be undeniable, as colonized plants are the last to show the impact of drought compared to controls. Before and after pictures speak a thousand words.

Soil and nutrient runoff from an Iowa farm field after rain.

THE IMPORTANCE OF THE HOST PLANT

So far, no one has been able to grow arbuscular fungi successfully without the presence of a host plant root or its exudates, though a few fungi in the wild germinate without exudates. Because arbuscular fungi are so important to restoration, horticulture, and agriculture, many attempts have been made to grow them in the lab without a plant host, but without success (although they have been grown in the lab with plant exudates added).

Studies have also shown that the presence of arbuscular mycorrhizae can improve the ability of host plants to withstand lower temperatures. We know that these associations lower the impact of stress on plants as a result of increased growth and more extensive root system, and these factors may contribute to temperature tolerance as well. Experiments with *Claroideoglomus etunicatum* and corn, for example, showed lower amounts of carbon dioxide stored between plant cells and higher stomatal conductance, which impact the tolerance of water-filled leaves and

In laboratory testing, tomatoes grown with mycorrhizal associations (left) were more healthy than those grown without (right) in drought conditions.

vascular parts to cold. (Perhaps someday I can use the help of mycorrhizal fungi to grow oranges in Alaska.)

WEED CONTROL

Some plants that are considered noxious weeds in agricultural and gardening environments also form arbuscular mycorrhizae, including the common dandelion (*Taraxacum officinale*), green foxtail (*Setaria viridis*), wild oat (*Avena fatua*), Canada thistle (*Cirsium arvense*), and chickweed (*Stellaria media*). Many prolific weeds do not form mycorrhizae, however, particularly those in the families Amaranthaceae, Chenopodiaceae, Brassicaceae, and Polygonaceae. These include wild buckwheat (*Fallopia convolvulus*), lamb's quarters (*Chenopodium album*), stinkweed (*Thlaspi arvense*), redroot pigweed (*Amaranthus retroflexus*), and kochia (*Bassia scoparia*).

Scientists are studying the effects of certain arbuscular mycorrhizae on weeds. Arbuscular mycorrhizal plants compete effectively and can starve nonmycorrhizal weeds of phosphorus, thus reducing their populations. In addition, some weeds are negatively impacted by the addition of certain mycorrhizal fungi. Imagine adding a mycorrhizal fungus that kills dandelions in your lawn, or planting a mycorrhizal

Potato production suffered during drought without mycorrhizal fungi associations (left); production was improved when mycorrhizae were present (right).

host plant whose fungal partner not only provides it with nutrients but also kills nearby chickweed. These types of commercial applications may become possible, along with others. For example, scientists could find a way to add strigolactones, the hormones produced by plants that stimulate the branching and growth of mycorrhizae, to soil to cause plants to form mycorrhizae quicker. Or perhaps they could manipulate plants to restrict strigolactone production to prevent particular arbuscular mycorrhizal formation, resulting in weak weeds that die.

AGRICULTURAL PRACTICES AND MYCORRHIZAE

Many modern agricultural practices are detrimental to arbuscular mycorrhizal fungi. As soil disturbances knock out mycorrhizal populations, weeds compete with crops for phosphorus; if the necessary mycorrhizal populations are not present, crop plants can suffer. Because the fragile arbuscular mycorrhizal fungi generally occur in the top 6 to 15 inches (15 to 38 centimeters) of the soil, crop rotation, fallowing, tilling, fertilizing, and soil sterilization can negatively impact mycorrhizal establishment and operation.

Dandelions are one of the many weeds that form arbuscular mycorrhizae.

MYCORRHIZAL FUNGI AND ALLELOCHEMICALS

Some mycorrhizal fungi are negatively impacted by the allelochemicals released by certain plants. Some of these can prevent the germination of the fungi, and others prevent the formation of mycorrhizae. Consider a case in Minnesota, where lakeside maple trees are disappearing because of the introduction of a nonnative species. Fishermen toss worms into the forest after spending the day fishing. The worms are not regular European earthworms, but Asian earthworms that reduce the duff too quickly, resulting in bare soils that provide space for garlic mustard (*Alliaria petiolata*), an invasive species, to grow. The garlic mustard exudates include allelopathic chemicals that kill the mycorrhizal symbionts of maple trees.

Rotating crops

Crop rotation is a standard practice in agriculture and gardening and is recognized as an effective system for improving and maintaining soil quality. However, the type of crop planted can have a profound effect on whether arbuscular mycorrhizal fungi can survive. Planting mycorrhizal host plants increases the ability of the soil to establish or maintain mycorrhizal fungi, but studies have shown that the opposite is also true: if no host plants are present, the fungi cannot get the carbon they need to survive. And with few or no spores remaining in the soil, reestablishing arbuscular mycorrhizal fungi through inoculation can take up to two months. Unfortunately, not every farmer or gardener can afford to wait that long, especially in areas with short growing seasons (such as Alaska, where I live).

Crop plants differ in their dependence on arbuscular mycorrhizal fungi. Corn, flax, legumes, and potatoes benefit significantly from arbuscular mycorrhizae, while wheat, oats, and barley benefit somewhat. Some crop plants, such as brassicas (mustards, cabbages, and broccoli, for example), beets (*Beta vulgaris*), and buckwheat (*Fagopyrum esculentum*), do not form mycorrhizae at all, or, like mustards, rapeseed, and radishes, release allelopathic chemicals when they decompose, which may prevent the formation or propagation of arbuscular mycorrhizal fungal spores (and sometimes other plants). These chemicals can remain in the soil even when the plants are removed.

Fallowing

Leaving a field fallow (cropless) can reduce the population of arbuscular mycorrhizal fungi in the soil because no host plants are present to provide the carbon the fungi need to survive. In addition, fallow fields are usually plowed or tilled, sometimes several times in a single season, which destroys the mycorrhizal network and moves fungal spores outside the growing area. If a field is not producing crops, it should be planted with a viable mycorrhizal host plant as a ground cover. If a field must be left fallow instead of being planted with a host crop, it should be inoculated with mycorrhizal fungi prior to the next growing season.

Disturbing the soil

Any surface disturbance of plants and soil, such as stock grazing, plowing, or tilling, can result in negative impacts to the growing environment, including the belowground mycorrhizal network. If the supply of carbon is reduced as a result of damage to the host plants or surrounding soils, mycorrhizae can be inhibited or destroyed. In addition, endophytic fungi (harmless, nonsymbiotic fungi present in every plant) in the leaves may be activated by damage and can affect mycorrhizae.

Soil disturbances result in damage to and loss of the soil and the mycorrhizal network.

Because arbuscular mycorrhizal spores are abundant in the top layer of soil, when disturbances such as tilling occur, spores are often destroyed or displaced to deeper zones, where they may not be reachable by roots or root exudates, or their germination and growth may be delayed because of unsatisfactory soil conditions. Hyphae are also destroyed by excessive soil disturbance, especially if tilling occurs during the fall. If the fungal hyphae are damaged or destroyed, they will not be available to partner with cover crops or crops planted in the spring, and the amount of phosphorus available for the next crop is reduced until a new network is formed.

Applying chemicals

Some fungicides, inoculants, fumigants, and fertilizers—especially those with high phosphorus content—can destroy or reduce the formation of arbuscular mycorrhizae by impacting some aspect of the soil food web upon which they rely. Agricultural seed is often coated with fungicides, and fungicides are often used in agricultural soils. These biocides can be deadly to arbuscular mycorrhizal fungi because they attack a particular fungal component such as their unique cell walls or a metabolic process.

Timing and rate of application of fungicides may determine whether the chemicals will affect mycorrhizae. Spraying plant leaves, for example, would have far less impact than drenching the soil with fungicide as mycorrhizae are beginning to form. In addition, not all fungicides impact and damage arbuscular mycorrhizal fungi. For example, at least one class of fungicides contains metalaxyl, which stimulates the formation of some mycorrhizae, perhaps because of the elimination of competing fungi by the fungicide.

Although herbicides and pesticides are not usually toxic to mycorrhizal fungi, they do have nonselective reach and can thus impact them. Arbuscular mycorrhizal fungi are living organisms and part of the soil food web. If a mycorrhizal host plant is killed after an herbicide application, for example, the number of spores in the soil can be reduced, with varying detrimental impacts.

Some agricultural chemicals can impact arbuscular mycorrhizae in positive ways. Application of certain pesticides, for example, result in higher colonization by arbuscular mycorrhizal fungi. This may result from killing off the pest, which results in healthier plants that can produce more exudates and attract and support more fungi. Nevertheless, organic solutions are always preferred over chemical ones when it comes to mycorrhizae.

Applying fertilizers

The amount of phosphorus or nitrogen in the soil greatly affects the germination and formation of arbuscular mycorrhizal fungi. With too much of either nutrient in the soil, mycorrhizal spores are less likely to germinate and mycorrhizae growth is hindered.

To compensate for the unavailability of phosphorus, many farmers saturate soil with synthetic chemical phosphate fertilizers, because once all the soil exchange sites are full of phosphorus ions, any excess nutrients will be more readily available to the plants. This practice results in a huge excess of phosphorus in the soil, which inhibits mycorrhizae from forming, and it can take decades for phosphorus levels to reduce to the point where the arbuscular mycorrhizae can thrive again. Excess phosphorus can be problematic for traditional organic farmers as well, who amend soils with animal manures that contain high levels of phosphate salts, often more than 1000 parts per million (which is so high that under natural conditions, soil amended with manure will still result in excess phosphorus 100 years from application). The use of phosphate fertilizers, manure-laden compost, and direct use of manures must be carefully managed to realize and maintain the maximum benefits provided by arbuscular mycorrhizae.

It makes sense to ensure that arbuscular mycorrhizal fungi thrive in agricultural soils. Fertilizer costs, especially phosphorus, are rising because of supply constraints. Arbuscular mycorrhizal plants often grow two to three times larger if less fertilizer is used, and scientists have demonstrated that plants increase their phosphorus and nitrogen uptake when arbuscular mycorrhizal fungi are maintained.

Burning fields

Although arbuscular mycorrhizal spores can tolerate cold temperatures, as evident by their presence in environments with severe winters, excessive heat (above 120°F, or 49°C) will kill or damage them. Some farmers burn fields after harvest, and some very good soils have been created

ERICACEOUS PLANTS AND MYCORRHIZAE

Given the size of the market, a great deal of effort has gone into creating an appropriate mycorrhizal spore blend for ericaceous plants such as blueberries, cranberries, heathers, and rhododendrons; but despite many efforts, no one has yet been able to produce a viable commercial mycorrhizal spore mix for use with these plants. Some available products can be used to form ericaceous mycorrhizal fungi, however; they contain mixes of colonized roots and the soil in which they grew. These have been shown to be effective ericaceous mycorrhizal inoculants.

In particular, unsterilized peat moss that has previously supported ericaceous plants can be an effective medium for inoculating plants grown in it. A study from the University of Vermont used natural peat mosses to provide the necessary propagules to create ericaceous mycorrhizae. Plants grown in ten of the 13 commercial brands tested established the fungi. If you want to establish mycorrhizae in ericaceous plants in either an agricultural or a nursery setting, using peat (and perhaps other soils) taken from the root zone of naturally colonized plants seems to be the only way to go. This works best if parts of the plants' roots are included and if the harvesting occurs in the late summer or fall.

as a result of these burning techniques. This practice creates quite a bit of heat, however—enough to kill the spores of arbuscular mycorrhizal fungi, especially in the surface soil layer. After a field is burned, the soil can be inoculated to reestablish the arbuscular mycorrhizal fungal populations.

Using GM crops

Some crops have been genetically modified to release *Bacillus thuringiensis* (Bt), a bacterium that is toxic to many insects, into the rhizosphere at the same time that mycorrhizal fungi spores germinate. The Bt toxin apparently negatively impacts the helper organisms in the mycorrhizosphere and the mycorrhizal colonization of the plant roots and surrounding soil. As more genetically modified (GM) crops are released, studies continue to examine the relationship between them and mycorrhizal fungi.

BENEFICIAL AGRICULTURAL PRACTICES

Although some standard agricultural practices such as frequent tillage and heavy phosphorus fertilization negatively impact mycorrhizae, farmers can use sustainable practices to bolster native mycorrhizal fungal populations. Even soils that have been intensively managed for an extended period of time contain populations of mycorrhizal fungi that can be augmented by using organic amendments and mulches, appropriately rotating crop production, using cover crops, and growing crops that form a symbiosis with arbuscular mycorrhizal fungi.

Using organic amendments and mulches

Studies have shown that soils in an organically maintained farm carry the same arbuscular fungal colonization as nearby undisturbed land. Organic mulches such as chicken litter, rice hulls, straw, and sewage sludge can increase colonization of host plants, but each type of mulch has a unique impact on mycorrhizal colonization and reproduction rates.

Composts and composted animal manures are also compatible with mycorrhizal fungi. Still, the variabilities of these require testing, not so much because of fear of negative results (though heavy and frequent use of manures can result in a buildup of phosphorus and inhibit mycorrhizal colonization), but to perfect their use. All organic amendments and mulches provide a complete set of soil food web organisms that break down the matter and enable mycorrhizal fungi to take advantage of nutrient mineralization. The fungi's acids can be directly involved in this decay.

Testing for and introducing mycorrhizal fungi

A successful farmer, like any good businessperson, evaluates change according to costs versus benefits. More and more studies show that using mycorrhizal fungi is a viable way to increase farm production and that they are worth the costs. Change can be slow, but with each passing year, the positive results of large-scale field testing are encouraging more farmers to consider substituting (or at least supplementing) the use of mycorrhizal fungi for chemical fertilizers to decrease costs and increase profits.

The level of existing arbuscular mycorrhizal fungi in soils can be determined by having the soil tested in a lab. Many labs provide results that indicate not only the presence or absence of mycorrhizal spores and propagules but the potential for mycorrhizae—soils with the appropriate chemical conditions. Specific plants can also be tested to determine whether they are associated with mycorrhizal fungi. Mycorrhizal activity is determined by sampling fine roots and then clearing and staining them to determine the percentage of the root length that contains mycorrhizal structures.

If conditions warrant, arbuscular mycorrhizal fungi can be introduced and plant roots colonized. The appropriate inoculum for each plant can be added using several systems, including drip irrigation, seed spreaders, rolled seed, and liquid spray.

USING INOCULANTS

There is no single perfect method for applying arbuscular mycorrhizal spores. Experiments with planting conditions, plants, and particular

Company
Field Way 0000
USA **1-Jan-15** **Report:15-0-0-0**

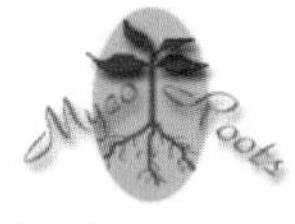

MycoRoots
Efren Cazares – Mycology Consultant, PhD
1970 NW Lance Way
Corvallis, OR USA. 97330-2209
541 752 0339
mycoroots@comcast.net
http://www.mycoroots.com

ARBUSCULAR MYCORRHIZAL FUNGAL SPORE EXTRACTION AND COUNT
Method: One sample of 20 gr dry weight of soil was processed by wet sieving/sucrose centrifugation.
Sieves used were 1 mm, 250 and 38 microns opening mesh respectively.
Analysis was done with a dissecting scope using 6-40x magnification.
ARBUSCULAR MYCORRHIZAL COLONIZATION
Method: Roots samples collected from each bag provided were cleared with KOH 10% and HCL 1%.
Staining was done with Trypan blue 0.5%.
Analysis was done with a dissecting scope using 6-40x magnification.

Mycorrhizal Assessment					
RESULTS	**SOIL**			**GRASS**	
	Arbuscular Mycorrhizal Fungal Spores			Roots	
Sample ID	**In 20 gr (dry wt)**	**In 1 gr.**	**Root length (cm)**	**Mycorrhizal colonization (cm)**	**Mycorrhizae %**
Field 1	404	20	200	21	11
Field 2	238	12	200	7	4
Field 3	546	27	200	45	23
Field 4	158	8	200	8	4
Field 5	132	7	200	7	4
Field 6	228	11	200	4	2
Field 7	162	8	200	4	2
Field 8	96	5	200	6	3
Field 9	14	1	200	6	3
Field 10	68	3	200	14	7
Field 11	180	9	200	12	6
Field 12	488	24	200	8	4

A report of a soil test shows the number of fungal spores, root lengths, and colonization of mycorrhizae in several samples.

SOME VEGETABLES AND HERBS THAT FORM ARBUSCULAR MYCORRHIZAE

basil
bean
borage
cannabis
caraway
carrot
celery
chive
corn
cucumber
dill
fennel
garlic
hop
hyssop
leek
marjoram
melon
mint
onion
oregano
paprika
parsley
parsnip
pea
pepper
potato
pumpkin
rosemary
sage
savory
tarragon
thyme
tomato
zucchini

SOME FRUITS THAT FORM ARBUSCULAR MYCORRHIZAE

apple
apricot
blackberry
blueberry
cherry
cranberry
currant
gooseberry
grape
morello cherry
peach
pear
raspberry
strawberry

SOME VEGETABLES THAT DO NOT FORM ARBUSCULAR MYCORRHIZAE

arugula
beet
broccoli
Brussels sprout
cabbage
cauliflower
cress
horseradish
mustard
radish
rapeseed
saltbush
spinach
turnip

spores have resulted in several inoculation schemes. The key is that the application be as efficient as possible: it must result in formation of mycorrhizae without wasting a lot of spores.

Most commercial mycorrhizal products include specific directions that should be followed. These inoculation methods may differ, but they all have one thing in common: as many of the fungal spores as possible must be located in a position that ensures that they will receive the host plant's chemical signals, which prompt the spores to branch their hyphal extensions and enter into mycorrhizal relationships with the roots. Without a signal from the plant, the mycorrhizal spore may germinate, but it will not colonize and it will not benefit the plant.

There are three types of formulations of arbuscular mycorrhizal propagules, a combination of spores and hyphal fragments that contain arbuscules, which can act like spores. All three formulations can comprise the same spores, but the delivery method is different.

Propagules mixed with granular substances These mixes are spread into soil or other planting media and onto roots in various ways.

Propagules mixed in soil These are natural soils that contain known mycorrhizal fungi. It is possible to make your own soil–fungi mix using plant root fragments.

Propagules mixed in liquid The liquid formulation is an important innovation that increases the uses of mycorrhizal fungal spores: they can be used in drip and irrigation systems and delivered to existing plants.

Formulations vary in the particular genus and species of included fungal spores. One size does not fit all. This is why it is vitally important that before applying the inoculant, you read the label and know what is included in the mixture. Make sure you are using the appropriate type of fungi to match the host plant, that the product is viable, and that the packaging offers a general idea of the numbers of spores involved—not so much for viability concerns, but for your own recordkeeping. You may have to experiment with various formulations and doses, so keeping good notes can help ensure that you settle on the best application methods and amounts.

The best way to establish arbuscular mycorrhizae in agricultural situations is to inoculate seed just before planting. This ensures that the plants can take advantage of mycorrhizal benefits from the first day after germination. Inoculant application methods vary depending on the situation at hand.

Seed inoculation Seeds can be rolled in granular and powdered formulations or sprayed with liquid formulas. Spraying seed with water first will help granular mixes adhere.

Seed germination mix inoculation All formulations can be used in the soil in which seeds are germinated. Seed-starting cubes, transplant cubes, and growing media for hydroponics can all be inoculated with granular or liquid formulations. Because mycorrhizal fungi thrive in a pH range of 5.5 to 7.0, the media used must have the proper pH to sustain the inoculant.

Bare-root inoculation Bare plant roots can be sprayed, sprinkled with, or dipped into any of the formulations and inoculation will result.

Bare-root soil inoculation Planting bare-root plants in soil that already contains arbuscular mycorrhizal propagules will result in the formation of mycorrhizae.

Transplant inoculation Both liquid and granular formulations can be applied to the plant roots at transplanting time. In addition, spores can be mixed into transplanting soil using any of the formulations.

Root inoculation All formulations can be used to feed roots of existing plants, as long as they contact the roots. Liquid formulations can usually be applied to the surface. Granular and soil mixes can be placed on the surface for plants with shallow roots in loose soils, but for deeper roots it is best to provide channels to the root zone. Packing these with granular or soil mixes allows for the root-to-fungi signaling required for the formation of arbuscular mycorrhizae.

MYCORRHIZAL CROP STUDIES

In 1902, in a pioneering study, G. N. Wissotzky demonstrated that areas of treeless Russian steppe (prairie) could support forest trees by adding mycorrhizal fungi taken from forest soils. This was the first successful use of a soil transfer of mycorrhizal fungi from one area to another. It showed the importance of the appropriate association between mycorrhizal fungi and host plant.

Today, after years of studies in which plants have been paired with arbuscular mycorrhizal fungi, we know which arbuscular mycorrhizal fungi or mixes are best used with a particular host plant. As a result, many companies supply propagules and can offer advice with regard to which fungi to apply as well as how to apply and maintain it. Always read labels carefully to determine whether a product will be appropriate for the plants being grown.

A variety of crops have been paired with arbuscular fungi in various studies. Several species of fungi are often used in these studies; look for them on the labels of commercial products. In all these studies, control plants were not inoculated.

Alfalfa (*Medicago sativa*) Studies with alfalfa, a legume whose roots form nitrogen-fixing nodules, show that inoculated plants grow larger root masses and form more nodules than uninoculated plants. In addition, alfalfa plants grown with seeds inoculated with *Funneliformis mosseae* fungi developed twice the number of arbuscules and were almost twice the biomass of control plants. Inoculation with *Rhizophagus fasciculatus* and *F. mosseae* resulted in reduced incidence of both *Verticillium* and *Fusarium* wilts, with lower numbers of the wilt-causing spores remaining

A view of Konza Prairie, an 8600-acre (3487-hectare) preserve of native tallgrass prairie in northeastern Kansas. Because the soils do not support the appropriate mycorrhizal fungi, trees do not grow on this or any prairie.

in the soil. Inoculated plants also took up more phosphorus than control plants.

Asparagus (*Asparagus officinalis*) When inoculated with several species of arbuscular mycorrhizal fungi, asparagus plants grew taller, with more crowns and shoots than control plants. In particular, asparagus inoculated with *Glomus* R10 and *Rhizophagus intraradices* gained more mass, with twice the survival rate of the control plants after 14 months. Plants associated with *Glomus* R10 experienced fewer incidents of disease. Increased levels of phosphorus and other nutrients resulted from inoculation with *Diversispora versiformis*, and those inoculated with *Gigaspora margarita* and *Claroideoglomus etunicatum* also showed increased growth and increased nutrient uptake.

Avocado (*Persea* spp.) Micropropagated plants inoculated with *Glomus deserticola* and *Funneliformis mosseae* showed better development and longevity than control plants. Control plant growth rate was slow when compared to plants inoculated with *Scutellospora heterogama*, *Acaulospora scrobiculata*, *Claroideoglomus etunicatum*, and *Rhizophagus clarus* because of increased nutrient uptake in the inoculated plants.

Banana (*Musa* spp.) The root systems of bananas inoculated with *Rhizophagus intraradices* showed less impact after infection by nematodes (*Radopholus citrophilus*) than a control group.

Barley (*Hordeum vulgare*) Barley inoculated with *Funneliformis mosseae*, *Rhizophagus fasciculatus*, and *Gigaspora margarita* increased seed yield by 27 percent and seed phosphorus content by 35 percent.

Barley inoculated with mycorrhizal fungi (right) shows better root growth than uninoculated plants.

Basil (*Ocimum basilicum*) Inoculated basil seedlings showed growth increases of 400 percent after they were transplanted to pots in a greenhouse. Survival rates were greater as well, with leaves showing the ability to adjust to drought. One study reported using *Gigaspora margarita*, *Rhizophagus clarus*, and 13 other mycorrhizal fungal species that had been isolated from growing basil.

Cannabis (*Cannabis sativa*) *Rhizophagus intraradices* and *Funneliformis mosseae* are advised for growing hemp and cannabis, with *F. mosseae* shown to increase growth. At least one study showed an increased number and improved size of flowers produced. Nutrient uptake was also increased in inoculated plants. Some studies indicated that after inoculating seeds, full mycelial colonization could take up to six weeks; however, plants benefitted prior to the six weeks as colonies became established.

Chickpea (*Cicer arietinum*) Colonization with *Funneliformis mosseae*, *Rhizophagus intraradices*, *R. fasciculatus*, and *Scutellospora gilmorei* reduced root galling and nematode populations in the soil surrounding chickpeas. *Funneliformis mosseae* and *R. fasciculatus* were the most effective. Inoculation with *R. intraradices* improved growth.

Citrus (*Citrus* spp.) When citrus plants were inoculated with *Rhizophagus fasciculatus*, plants were healthier and weighed more than control plants, with improved nutrient uptake. Studies showed that mycorrhizal inoculation reduced the density of root nematodes.

Corn (*Zea mays*) Corn root tissues increased by 35 percent with inoculation of *Funneliformis mosseae* and 98 percent with inoculation of *Rhizophagus fasciculatus*. After inoculation with *R. intraradices*, drought wilting was

Corn roots inoculated with mycorrhizal propagules (right) grew thicker than those that were uninoculated.

delayed and plants better withstood salt stresses. Mycorrhizal hyphae bring more nitrogen to roots during drought, and this probably helps with increasing drought tolerance.

Cotton (*Gossypium hirsutum*) *Funneliformis mosseae* and *Rhizophagus intraradices* form mycorrhizal associations with cotton. In one study, ammonium and nitrates in the soil of inoculated plants were lower, suggesting that the inoculated plants absorbed more of these nutrients than the control plants. In addition, root galls and nematode populations were reduced after mycorrhizal colonization.

Cucumber (*Cucumis sativus*) *Funneliformis mosseae* and *Rhizophagus intraradices* were paired with cucumbers, resulting in better toleration of salt stress. Phosphorus uptake was improved after inoculation and may have contributed to larger cucumbers.

Eggplant (*Solanum melongena*) Mycorrhizal colonization protects eggplant against *Verticillium* wilt, or at least delays its impact. Plants inoculated with *Claroideoglomus etunicatum* fared better than those with *Gigaspora margarita*, yielding more and larger fruit.

Ginger (*Zingiber officinale*) Nematode populations were reduced in plants inoculated with *Rhizophagus fasciculatus* (especially) and *Funneliformis mosseae*.

Grape (*Vitis* spp.) Inoculated vines showed increases in dry weight, height, and nutrient uptake. Inoculation with *Rhizophagus fasciculatus*, *Funneliformis mosseae*, and *Glomus macrocarpum* resulted in significant increases in yields and improved ability to sustain drought stress. Grapevines inoculated with *G. deserticola* showed earlier bud and flower emergence, fruit set, and ripening.

Hops (*Humulus lupulus*) *Claroideoglomus etunicatum* was effective in colonizing hops. Shoot dry weight was increased with inoculation, but root dry weight was not.

Lettuce (*Lactuca sativa*) *Funneliformis mosseae*, *Glomus deserticola*, and *Rhizophagus fasciculatus* all colonize lettuce, but *R. fasciculatus* helped particularly with salt stress. Lettuces take up more nitrogen when associated with mycorrhizae. Association with *R. fasciculatus* improved phosphorus uptake, and *F. mosseae* improved nitrogen uptake in drought conditions, suggesting a mixture is best.

Mint (*Mentha arvensis*) Colonization with *Gigaspora margarita* and *Rhizophagus clarus* greatly increased the size of mint plants when phosphorus was lacking. *Acaulospora scrobiculata* seemed to improve plants even more with just a bit of phosphorus added. Essential oils and menthol levels were lower in plants that were not colonized. Colonization with *R. intraradices* greatly increased plant height and leaves.

Oat (*Avena sativa*) Inoculation with *Glomus* T6 and D13 in one study resulted in as much as 70 percent root and 55 percent shoot growth increases. Translocation of heavy metals from soil to shoots in roots was less when plants were colonized. Phosphorus uptake was not increased, however.

Okra (*Abelmoschus esculentus*) Colonization with *Funneliformis mosseae* and *Rhizophagus fasciculatus* reduced nematode populations; *F. mosseae* was the more effective of the two.

Onion (*Allium cepa*) Inoculation resulted in improved uptake of phosphorus, and in tests in low-phosphorus soils, uninoculated plants were stunted. Inoculated plants also fared much better under salt stress.

Peanut (*Arachis hypogaea*) Plants inoculated with *Gigaspora margarita* took up six times more phosphorus than uninoculated plants. An increased tolerance to nematode attacks was shown in mycorrhizal plants. *Funneliformis mosseae* inoculation resulted in better biomass and less root and pod rot.

The inoculated lettuce (right) grew larger than the uninoculated plant.

Pepper (*Capsicum annuum*) *Rhizophagus fasciculatus* was generally more effective than *Glomus* ZAC-19 in increasing growth of colonized plants, especially under reduced phosphorus conditions.

Potato (*Solanum tuberosum*) The arbuscular mycorrhizal fungi associated with potatoes in their native Andean ecosystems is altitude affected, and the highest altitude has the greatest species diversity. In the Andes, various species from eight of the 11 Glomeromycota families colonize potatoes. The three most common are *Funneliformis mosseae*, *Rhizophagus irregularis*, and an unidentified *Claroideoglomus* species.

Inoculation increased the number and size of tubers; one study showed an almost 50 percent increase in number. Studies indicated that varieties show selectivity regarding the type of fungal species to associate with. Mortality tests of plants colonized by the pathogenic fungus *Rhizoctonia* showed that *Claroideoglomus etunicatum* and *Rhizophagus intraradices* impacted Gold Rush and LP899221 varieties differently. *Rhizophagus intraradices* produced the most potatoes in one test. Mycorrhizal fungi strains should be tested for the particular variety of potato.

Raspberry (*Rubus* spp.) and strawberry (*Fragaria* spp.) *Rhizophagus intraradices* colonization of potted strawberry plants transplanted to the field resulted in healthier plants with more runners and greater biomass.

More and bigger potatoes (right) are produced after inoculation with mycorrhizal fungi.

On raspberries, *Gigaspora* species colonization depressed growth. In general, however, mycorrhizal colonization increased the ability of both plants to recover from drought, probably due to increased access to water through the expanded root system. Plants inoculated with *Funneliformis mosseae* and *R. fasciculatus* showed reduced damage by weevils.

Soybean (*Glycine max*) Inoculation with *Rhizophagus fasciculatus* positively impacted yield, seed weight, and pod and seed numbers. *Funneliformis mosseae*–inoculated plants better sustained drought conditions. Colonization seems to be slower in tilled soils, and there is less phosphorus uptake.

Tomato (*Lycopersicon esculentum*) Mycorrhizal inoculation increased tomato plant growth, height, and weight. Plants colonized with *Funneliformis mosseae* prior to being infested by nematodes survived, and mycorrhizae made up for the loss of roots resulting from the infestation. Colonization with *Rhizophagus fasciculatus* improved nutrient uptake and reduced the size of nematode galls; it also improved the plant's ability to tolerate drought and salt stress and resulted in more and larger fruit. One study showed that inoculation plus high phosphorus content in the soil resulted in fewer days until first flower and more flowers.

Watermelon (*Citrullus lanatus*) Inoculation with *Rhizophagus clarus* helped mitigate drought stress.

Wheat (*Triticum* spp.) Wheats are always dependent on mycorrhizal fungi. After inoculation, dry weight increases in one study ranged from 29 to 100 percent.

Inoculated strawberries (left) grow larger and are healthier than those without mycorrhizal associations.

Mycorrhizae in Horticulture

Most annual and perennial plants form arbuscular mycorrhizae, and many plants respond readily to inoculation. In fact, almost any nursery-grown plant or start forms arbuscular mycorrhizal associations when planted in the garden. Some plants have a difficult time thriving without mycorrhizae, especially trees and shrubs, and plants with thick, carrot-like roots. Again, plants that do not form mycorrhizal relationships are members of families Amaranthaceae, Brassicaceae, Caryophyllaceae, Chenopodiaceae, Polygonaceae, Portulacaceae, and Proteaceae.

Inoculating seed and/or soil with mycorrhizal fungi can be a cost-effective and conscientious way to minimize potentially polluting nutrients such as nitrogen and phosphorus in runoff water. Adding mycorrhizal propagules to soils greatly increases the uptake of water and mineral nutrients, and mycorrhizae can protect plant roots from attack by pathogens. Despite these and other practical benefits of mycorrhizal inoculation, however, many nursery-grown seedlings are planted in sterile media that is heavily fertilized.

Mycorrhizae do not establish in beneficial numbers if ample phosphorus is already available in the planting media. Nurseries often use premixes that contain enough phosphorus and other nutrients for the plant to grow at the nursery, and at home, for a while. Soils are sterilized and fumigated to prevent the spread of pathogens, and some of the peat used in planting mixes does not contain and support growth of the appropriate types of mycorrhizal fungi.

Special mycorrhizal formulations and delivery systems have been developed to make inoculation easier and more cost-effective in a nursery setting, and a great number of studies are confirming the benefits of inoculating nursery plants:

- Plants started from seed can be inoculated at the earliest growth stage, offering seeds and seedlings the advantages of early mycorrhizal association.
- For plants that are propagated using tissue culture, adding mycorrhizal propagules to the soil mix before planting can help the tiny cuttings establish strong root systems so healthy plants can be transplanted earlier.
- A plant with established mycorrhizal root associations has a stronger connection to the planting media and more readily accepts nutrients. It can also handle stresses such as lack of water and excessive heat.
- Plants with mycorrhizal associations can better withstand damage by common root pathogens, such as nematodes and mites.
- Seedlings grown with mycorrhizal fungi exhibit a uniformity that is often required in horticultural displays and landscaping beds. Because all seedlings in a flat can share the same mycorrhizal matrix, they can share nutrients and other benefits.
- Nursery inoculation helps ensure that strong plants will survive and thrive in their next location, resulting in fewer returns and increased customer satisfaction.

USING INOCULANTS

Both arbuscular and ectomycorrhizal fungi inoculants are available to nursery wholesalers, retail nurseries, and home growers. (You can also make your own inoculants, as you'll learn later in this book.) Two general types of commercial inoculants are available: a grab bag of fungal species that may include other kinds of microbes, and propagules of one specific kind of mycorrhizal fungus.

Most blends of mycorrhizal fungi are available as spores or hyphal fragments mixed with various media such as sand, peat, clay, gel, or water to ensure delivery to the plant's root system. Many potting mixes and composts also add mycorrhizal fungi to their formulations. A list of mycorrhizal species and their concentrations is usually included on the packaging of products containing propagules. Some manufacturers list them simply as propagules, while others add more specific information. It is always a good idea to read the package ingredients list to identify the type of fungi (endo- or ectomycorrhizal) and the species included, so you can match the fungi to the plant. Also useful is the percentage of spores and/or propagules that will germinate.

In some countries there is governmental regulation of commercial formulations of mycorrhizal fungi. In the United States, for example, spore counts and identification of the mycorrhizal fungi in the formulation must be included on the product package. This information says nothing of the fungi's viability, however, which can be affected by the age of the propagules and how they have been handled. Although extreme cold and some heat does not affect their viability, extreme heat can kill propagules, and moisture can trigger premature germination (which proves that even fungi can experience dysfunction). In general, you can rely on viable spores lasting a year or two and root and hyphal fragments to last a year or more.

How much to use? A fungus can regenerate from just a tiny amount of a hypha. Determining the appropriate number of propagules to apply is not easy, though some manufacturers do offer recommendations. It depends on the plant, the fungi, the soil conditions, the delivery media, and other factors. Experiment and review the literature. Fortunately, it seems that you cannot apply too many mycorrhizal propagules to plants. Different nursery conditions mandate different usage depending on fertilizer and pesticide application practices. Nurseries can inoculate seed and transplants with liquid and granular mixes, and liquid formulations can be added to drip systems.

Timing the inoculation

The timing of inoculation is important to the fungi's effectiveness. Apply the inoculant too late, and its benefits are lost. It is best to inoculate as early as possible in the plant's life to ensure that it gets the maximum benefits from the mycorrhizal association. Roll or spray seeds in mycorrhizal formulations before they germinate.

Establishing mycorrhizae in planted containers can take up to two months—a lot less time than it can take to colonize an agricultural field. This is because the conditions in pots are usually far more ideal and controllable than those in the field, and propagules are applied in much greater concentrations in pots. Container-grown mycorrhizal fungi also produce spores much more quickly. Sporulation is usually triggered by crowding in the pot as mycorrhizae develop. As soon as the pot becomes full of fungi, they begin the reproduction process. Container-grown mycorrhizal fungi can begin reproducing via spores within a year of inoculation.

Placing the propagules

Mycorrhizal fungi spores will not grow and thrive and hyphal fragments will not develop if they do not come in contact with the proper root exudates. To ensure proper propagule placement, make sure propagules are in physical contact with plant roots. The standard advice is to roll the roots of transplants in mycorrhizal mixes or sprinkle the mix directly on exposed roots. Once the roots are inoculated and the plant is transplanted, the extracellular hyphae grow from the roots and into the soil in the container.

Studies have shown that it is also good practice to mix propagules throughout the potting mix before transplanting. The spores in the soil will germinate, grow, and increase, and more mycorrhizae will form as they become exposed to the expanding network of roots. A larger mycorrhizal network results, with increased nutrient uptake and other benefits. As plants are transplanted to larger pots, the existing fungal network comes along, though adding more fungal spores to the transplant mix continues to increase the number of mycorrhizae.

FERTILIZERS AND MYCORRHIZAE

As more nurseries sell stock with mycorrhizae, gardeners must be sure not to destroy the fungi. One way is to avoid overfertilizing. Too much phosphorus or nitrogen won't necessarily kill off the mycorrhizal fungi, but it will slow or limit their growth and colonization of plant roots. In general, mycorrhizal fungi improve plant growth in low-nutrient soils, especially those with limited phosphorus. As more fertilizer is added, the mycorrhizal colonization generally declines. As the fungal colonies decrease, the concentration of phosphorus in plant tissues also decreases.

Studies indicate that organic fertilizers are generally compatible with mycorrhizae, whereas phosphorus-rich inorganic fertilizers can inhibit mycorrhizal growth. Most fertilizers are packaged and labeled to guarantee a certain percentage of nitrogen, phosphorus, and potassium (N-P-K) in the mix. Most gardeners think the N-P-K numbers represent the percentages by weight of the three nutrients, but that's not entirely true. The phosphorus and potassium numbers on the label represent the percentage of phosphorus pentoxide (P_2O_5) and potassium oxide (K_2O), respectively. To make the conversions for a pure N-P-K trilogy, you need to multiply the number for phosphorus by 0.44 and that for potassium by 0.83 to determine the actual weights of phosphorus and potassium.

Some experts recommend that to establish and maintain arbuscular mycorrhizae, less than 80 parts per million of phosphorus or fewer than 75 micrograms of phosphorus should be present per gram of soil or growing media. These numbers are in contention, however, as some plants readily form arbuscular mycorrhizae even in the presence of high amounts of phosphorus; studies are being conducted to arrive at new figures, including figures for specific plant types.

Finally, when potted plants are transferred to the garden, gardeners can again supplement the mycorrhizal fungi at the grow location to bolster the plant and surrounding soils.

USING POTTING AND COMPOST MIXES

Growing mixes are critical to forming and sustaining arbuscular mycorrhizae. Some manufacturers of commercial potting and growing mixes sterilize their growing media, which renders them free from microbes, including mycorrhizal fungi. Other manufacturers add mycorrhizal propagules to their blends (quite a change after years of using mix sterility as a selling point).

Compost is free of mycorrhizal spores, because it contains no live roots to support mycorrhizae and because the temperatures used to produce compost kill mycorrhizal propagules. In addition, manures included in compost mixes can have very high levels of phosphorus. It is crucial that the levels of phosphorus be low enough in compost not to discourage the germination of inoculum and formation of mycorrhizae.

The best soil and compost blends ensure that pathogens will not prevail and that proper nutrients and microbes will be available to the plant and the mycorrhizal fungi. The needed microbes, along with mycorrhizal fungi, include nitrogen-fixing bacteria and phosphate solubilizers that produce the requisite hormones and control plant diseases. If mycorrhizal propagules have not been added to the mix, add them at planting time.

MYCORRHIZAL PLANT STUDIES

Thousands of studies on the horticultural uses of mycorrhizal fungi have been conducted by commercial entities, academic institutions, and home growers. The results of some of these are described here.

Inoculated daffodils (left) showed more growth and vitality than uninoculated plants.

Cyclamen (*Cyclamen* spp.) Inoculated plants showed better resistance to infection by the fungal pathogen *Cryptocline cyclaminis*. Mortality rates decreased in inoculated plants.

Daffodil (*Narcissus* spp.) Bulbs inoculated with *Funneliformis mosseae* showed improved flower yield, stalk length, and quality.

Freesia (*Freesia* spp.) Inoculated plants produced larger daughter corms and more flowers. Plants also grew roots and shoots more quickly.

Geranium (*Pelargonium* spp.) Plants inoculated with *Funneliformis mosseae* and *Rhizophagus fasciculatus* were better able to tolerate drought stress.

Gerbera (*Gerbera jamesonii*) A mix of *Rhizophagus intraradices* and *R. vesiculiferus* increased leaf and root dry weight, and inoculated plants flowered significantly earlier than the controls.

Hosta (*Hosta* spp.) Use of commercial mycorrhizal fungi resulted in greater top growth, an important development for commercial growers as well as home gardeners.

Marigold (*Tagetes erecta*) and zinnia (*Zinnia elegans*) Seeds inoculated with *Claroideoglomus etunicatum* flowered faster and produced more flowers.

Petunia (*Petunia* ×*atkinsiana*) Inoculated plants had higher reproductive growth than nonmycorrhizal control plants, with a threefold increase in vegetative growth, and colonized plants flowered 15 days

Inoculated marigolds (right) respond with more growth.

earlier than controls. In petunias colonized by *Rhizophagus irregularis*, nitrogen starvation resulted in formation of mycorrhizae even with a high availability of phosphorus, suggesting that mycorrhizae may form even when they are limited by nitrogen or phosphorus.

Poinsettia (*Euphorbia pulcherrima*) Cuttings showed improved growth as a result of inoculation with *Gigaspora margarita* and after adding spores to the plant misting system and potting mix.

Rose (*Rosa* spp.) *Rhizophagus intraradices* improved the growth of container-grown mini roses. For some varieties, using mycorrhizal fungi was as good as or better than using rooting hormone.

Snapdragon (*Antirrhinum* spp.) Plants inoculated with *Claroideoglomus etunicatum* outperformed controls with regard to size and blooms.

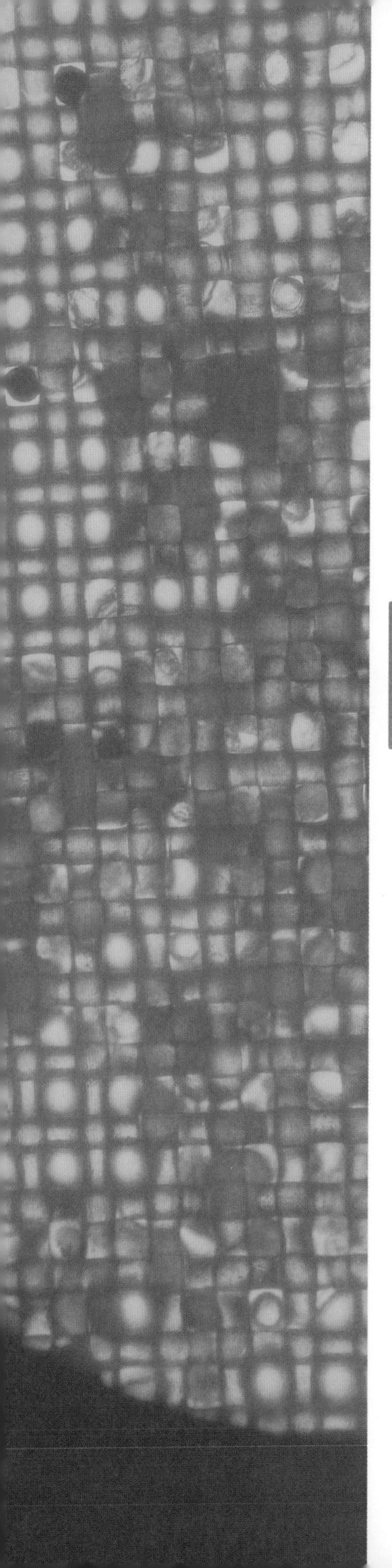

Mycorrhizae in Silviculture

Nearly every tree species forms some type of mycorrhizal association. Some trees form arbuscular mycorrhizae, but others have evolved over time and are hosts to ectomycorrhizal fungi. Some trees associate with both forms of mycorrhizae, though usually at different periods of their lives. Members of the pine family (Pinaceae), for example, are mostly ectomycorrhizal hosts, but they can also be colonized by arbuscular fungi during the early phases of establishment, especially in disturbed sites where arbuscular fungal propagules are more abundant than ectomycorrhizal fungi.

Although arbuscular mycorrhizal fungi partner with many trees, and they are included in this discussion, when appropriate, the emphasis here is on ectomycorrhizal fungi, mostly members of phyla Basidiomycota and Ascomycota, with a few from Zygomycota. About 6000 species of ectomycorrhizal fungi associate with about 10 percent of plant families, including Pinaceae (pines and other conifers), Cupressaceae (cedars and cypress), Fagaceae (beeches and oaks), Betulaceae (birches and alders), Salicaceae (poplars and willows), Dipterocarpaceae (lowland rainforest trees), and Myrtaceae (myrtle, clove, guava, and eucalyptus). The plants and trees they support are large in size, grow in groups, and cover huge portions of the Earth.

The number and density of the mycorrhizal colonies formed is staggering, as many trees and shrubs associate with more than one species of mycorrhizal fungi and often more than one type. In one Swedish forest, for example, biologists discovered that 95 percent of all the root tips had mycorrhizal associations and as many as 1.2 million ectomycorrhizae were found in a mere 1 square meter of soil.

The benefits of ectomycorrhizae to trees and shrubs are, for the most part, the same as those imparted by endomycorrhizae formed by arbuscular mycorrhizal fungi and roots (some of which colonize tree and shrub roots). A quick review of what they do, however, doesn't hurt. They increase root formation and mass, nutrient uptake, water uptake and storage, and stress and drought tolerance; they decrease nutrient runoff, moderate pH, and improve pathogen resistance.

Mycorrhizal fungi and plant roots access the same nutrients in the soil and use the same mechanisms to take up nutrients, but the fungi can also access minerals by growing into rock and dissolving them. This enables the fungi to capture mineral nutrients for the host plant that the plant would never get without the mycorrhizal association.

The soils of this hardwood forest are teeming with mycorrhizae.

TREES AND SHRUBS THAT FORM ARBUSCULAR MYCORRHIZAE ONLY

blackthorn (*Prunus spinosa*)
boxwood (*Buxus sempervirens*)
buckthorn (*Rhamnus cathartica*)
elder (*Sambucus nigra*)
elm (*Ulmus glabra*)
gorse (*Ulex europaeus*)
horse chestnut (*Aesculus hippocastanum*)
privet (*Ligustrum vulgare*)
spindle tree (*Euonymus europaeus*)
yew (*Taxus baccata*)

TREES AND SHRUBS THAT FORM ECTOMYCORRHIZAE ONLY

black pine (*Pinus nigra*)
common lime (*Tilia ×vulgaris*)
Douglas fir (*Pseudotsuga menziesii*)
downy birch (*Betula pubescens*)
dwarf birch (*Betula nana*)
dwarf cherry (*Prunus cerasus*)
European beech (*Fagus sylvatica*)
European hornbeam (*Carpinus betulus*)
fir (*Abies* spp.)
hazelnut (*Corylus avellana*)
larch (*Larix* spp.)
large-leaved lime (*Tilia platyphyllos*)
maritime pine (*Pinus pinaster*)
Norway spruce (*Picea abies*)
Scots pine (*Pinus sylvestris*)
silver birch (*Betula pendula*)
Sitka spruce (*Picea sitchensis*)
sweet chestnut (*Castanea sativa*)
wild service tree (*Sorbus torminalis*)

TREES THAT FORM ECTOMYCORRHIZAE AND ARBUSCULAR MYCORRHIZAE

alder (*Alnus glutinosa*)
alder buckthorn (*Frangula alnus*)
ash (*Fraxinus excelsior*)
bird cherry (*Prunus padus*)
black locust (*Robinia pseudoacacia*)
English elm (*Ulmus procera*)
European crab apple (*Malus sylvestris*)
field maple (*Acer campestre*)
gray alder (*Alnus incana*)
hawthorn (*Crataegus monogyna*)
holly (*Ilex aquifolium*)
juniper (*Juniperus communis*)
littleleaf linden (*Tilia cordata*)
Midland hawthorn (*Crataegus laevigata*)
Norway maple (*Acer platanoides*)
poplar (*Populus* spp.)
red-berried elder (*Sambucus racemosa*)
rowan (*Sorbus aucuparia*)
sycamore (*Acer pseudoplatanus*)
walnut (*Juglans regia*)
whitebeam (*Sorbus aria*)
wild cherry (*Prunus avium*)
wild pear (*Pyrus communis* subsp. *pyraster*)
willow (*Salix* spp.)

Mycorrhizal fungi also protect plants from damage by heavy metals. The fungi take up and retain arsenic, copper, zinc, iron, lead, cadmium, nickel, mercury, chromium, and aluminum, preventing their accumulation in the soil and consequent damage to the host plant and others. Some fungal tissues can even filter radioactive materials such as cesium.

A mycorrhizal fungus normally uses about 15 percent of the plant's photosynthetic products. A 15-percent loss of energy from photosynthesis is a lot—imagine losing 15 percent of your energy and having to make energy to do it! In return, however, the plant can get 85 percent of its nitrogen through mycorrhizae. The transfer of carbon to the fungus and the cross-transfer of fungi-obtained nutrients to the plant host occur in the intercellular (apoplastic) space between the hypha and the plant cell wall. Molecules flow through the walls for transport across the respective organism's membranes.

ECTOMYCORRHIZAE AND FOREST LIFE

In managing a forest, a silviculturist focuses on the biggest trees, not only because they will be harvested for lumber, but also because they protect the health of the understory, provide shade, supply organic

Monterey pine (*Pinus radiata*) seedlings germinate after a fire and provide host plants for mycorrhizae.

debris, and more. These large mother trees are colonized by mycorrhizal fungi, and the resulting mycorrhizae support all manner of other organisms, including other plants. They share their mycelial network with seedlings and young trees in the undercanopy that are not able to compete for sunlight. They also share the benefits of their sunlight-gathering capacity with their progeny via the underground extracellular network.

When big trees die in a forest, much of their nutrients are returned to the fungal network for reuse. This reuse also includes the decaying leaves that fall year after year, as evidenced by the mycorrhizal fungi, vesicles and all, which are found inside dead leaves, especially those close to colonized root systems.

Unlike arbuscular fungal spores, ectomycorrhizal fungal spores can survive forest fires by forming spore banks that reside in the soil until the spores are needed. Many shrub species rapidly sprout after a fire and keep mycorrhizal populations fed until regenerating trees can plug into the mycelial network. This enables mycorrhizal regeneration to occur immediately after a fire instead of having to wait for spores to blow in from the next forest over.

HOW ECTOMYCORRHIZAE FORM

The formation of ectomycorrhizae involves a complicated but beautiful arrangement between fungus and plant. From 25 to 30 chemicals are produced by both participants, most of which will never be used

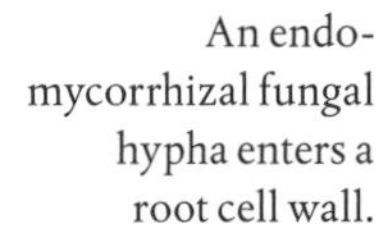

An endo-mycorrhizal fungal hypha enters a root cell wall.

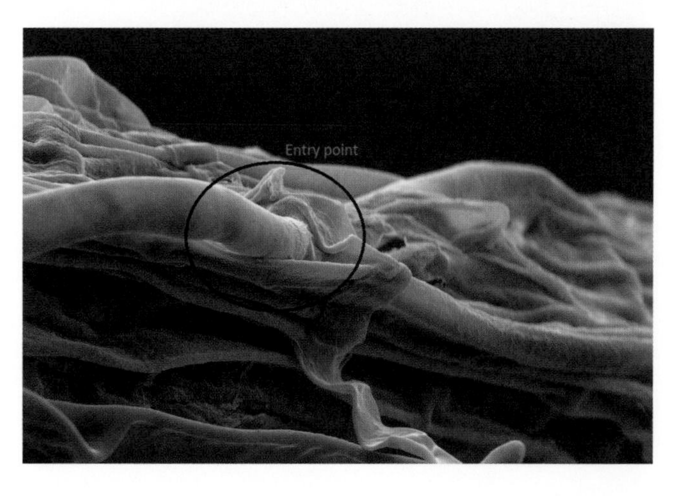

again for any other purpose, which emphasizes the importance of this process.

At the beginning of the process, the host plant roots send out signals via strigolactone hormones to encourage the mycorrhizal fungi in the area to extend their hyphal reach and seek out a partnership. When the growing ectomycorrhizal fungal hyphae reach the roots, they do not penetrate the cell wall as endomycorrhizal fungi do. Instead, they weave together to form a mantle around the root, and then they use cellulolytic enzymes to dissolve the middle lamellae (layer) of both epidermal and outer cortical cells. As a result, the hyphae surround the cell walls, forming the Hartig net. The plant cells are not penetrated, and the integrity of the protoplasm of each organism is kept totally intact. Within the Hartig net's massive surface area, the exchange of nutrients with the symbiont occurs.

In the mantle, the hyphal sheath that surrounds plant roots, carbohydrates from the host plant are stored and retrieved. The mantle formation slows or stops the development of root hair cells and forces the plant to rely more on the fungi to obtain its nutrients. Mantles can be up to 40 micrometers thick—so thick, in fact, that they completely cover the roots. After the mantle is created, the host plant's roots become short and thick, and their growth slows down.

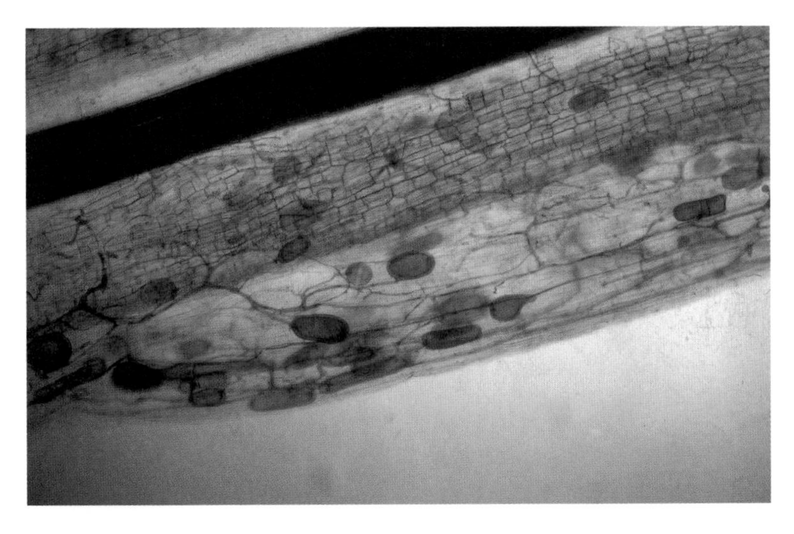

Stained tissues on both sides of the vesicles show the Hartig net created by the ectomycorrhizal fungus.

SQUIRRELY DEPENDENCIES

Interactions with other organisms are always a factor in the ability of ectomycorrhizae to form. Many microscopic soil food web organisms help support mycorrhizae, but help can come from other sources as well, including larger animals.

Kaibab squirrels (*Sciurus aberti* subsp. *kaibabensis*) live in the ponderosa pine forests on the north rim of Grand Canyon National Park and are found nowhere else in the world. The rodents are mycophagists—that is, they eat fungi—and in this case, they eat the truffles of a mycorrhizal fungus that grows on the roots of the pines. These sporocarps are full of water, nitrogen-fixing bacteria, yeast, and, of course, fungal spores. All this moves though the squirrel's digestive system and into its fecal pellets. Thus the squirrel becomes a major vector for distribution of the fungal spores. Without the squirrels and the surviving spores, there would be no ectomycorrhizae to support the trees in which the squirrels live.

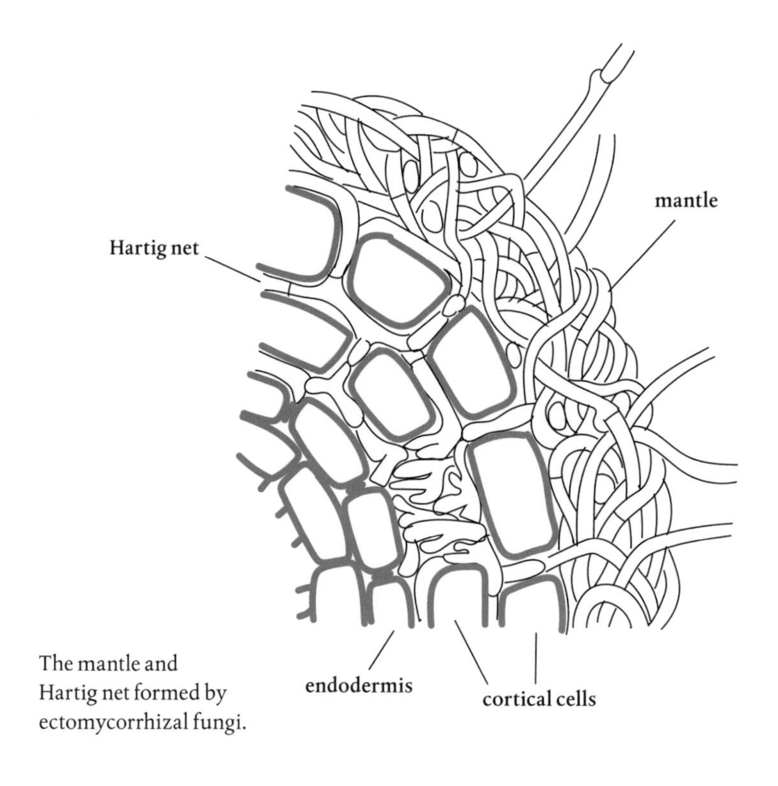

The mantle and Hartig net formed by ectomycorrhizal fungi.

Meanwhile, extraradical hyphae begin to develop in the soil. As with endomycorrhizal fungi, the ectomycorrhizal fungi excrete digestive enzymes and acids to help break down nutrients in the soil, which are transported through tunnels and channels in the plasma membrane. Cytoplasmic streaming in the fungal cell keeps things moving in the right direction.

Unlike arbuscular mycorrhizal fungi, most ectomycorrhizal fungi reproduce sexually, via large sporocarps, or fruiting bodies: mushrooms, puffballs, and truffles (though not all mushrooms and puffballs are ectomycorrhizal fungal fruits). Reproduction involves the transfer of genetic materials, development of spores, and then spore dispersal.

Many fruiting bodies of ectomycorrhizal fungi are safe to eat and have commercial value. Many are poisonous, however, such as the Alice-in-Wonderland toadstool *Amanita muscaria* (fly agaric), which associates with birch trees where I live in Alaska. Whether edible or not, all ectomycorrhizal fungal fruits produce small spores of usually less than 10 micrometers, which are dispersed primarily by wind. However, small mammals, birds, insects, slugs, worms, bacteria, and even other fungi

Trophies of a mushroom hunt: *Boletus edulis, B. reticulatus,* and *B. impolitus.*

eat spores and mushrooms or otherwise transport them. These ectomycorrhizal spores are more easily dispersed than the much larger and heavier endomycorrhizal fungal spores.

MATCHING FUNGI TO TREE

Some ectomycorrhizal tree hosts show specificity for particular fungal associations. The extent of this host specificity varies not only in the forest but in the lab as well. A set of chemical cues inform the plant that initiates the formation of mycorrhizae. Sometimes this is the only way that fungal spores will germinate. This protective measure has evolved to prevent other organisms from affecting germination, which would be to the detriment of both the fungi and plants, neither of which can survive without the other.

Some fungi are specialists and associate with only one host species or a limited number of plant hosts, while others are generalists and can associate with a wide range of hosts. In monocolonizations, one fungus invades one root. But an individual mycorrhizal fungus can be simultaneously associated with not only more than one host plant but with more than one species of plant. The host plant is also free to form associations with other organisms. A mature tree can be associated with ten or more different fungal species forming mycorrhizae. Things can get pretty complicated.

Modern techniques using DNA and RNA for identification enable scientists to determine the best host plant associations for particular mycorrhizae. A problem, however, is that some associations that occur in the lab do not occur in the wild, and some field-observed associations cannot be replicated in a lab setting. This makes matching tree to fungi very tricky.

Trees and shrubs without the proper mycorrhizal associations struggle and usually die because they cannot obtain enough nutrients and/or water. (A friend, for example, once planted 10,000 trees in Alaska without a mycorrhizal fungal inoculation. They all died.) A well-known example of the importance of these associations occurred in Puerto Rico in the 1950s, when attempts to grow pine trees using native island soils failed. The seedlings turned yellow because of lack of nutrients, even if extra fertilizers were added. Within two years after being planted, every tree died. In 1955, soils taken from around some pines in North Carolina were used to inoculate the roots of the pines on the island. These trees thrived because of the association with the appropriate ectomycorrhizal fungi.

Specific strains of ectomycorrhizal fungi, tested with specific hosts, are commercially available. Not only do these fungi help the plants grow in the nursery, the plants are healthier after being transplanted in the forest or field.

Good general-purpose ectomycorrhizal fungi for tree nurseries include *Thelephora terrestris*, *Laccaria laccata*, and species of *Inocybe*. These are found in natural forests and can be used to inoculate outdoor nursery-grown stock. Ectomycorrhizal *Pisolithus arhizus* and *Scleroderma citrinum* have been used for oak inoculation, and the latter is the best fungus for European beech. *Hebeloma arenosum* has been used on pine seedlings, and *Leccinum scabrum* is a good inoculant for birch seedlings. Arbuscular mycorrhizal *Rhizophagus intraradices* is used to inoculate seedlings of red cedar, redwood, and giant sequoia.

MYCORRHIZAL FUNGI IN THE FIELD

Silviculturists and arboriculturists can use mycorrhizal fungi to speed the growth process when reestablishing trees in disturbed areas, such as chemical spill sites, borrow pits, and heavily eroded terrain. Studies have shown that mycorrhizae also improve both growth and survival of seedlings in disturbed soils and during times of drought.

Reclamation of waste sites

Mycorrhizae can take up and isolate toxins—including heavy metals and radioactive elements. According to research, some mycorrhizal fungi bind and isolate radioactive elements, which can remain immobilized within the fungal tissue for years.

Consider several studies involving mushrooms related to the 1986 Chernobyl nuclear disaster. In 2002, a robot sent inside the destroyed nuclear power plant took samples of fungi growing on the walls. Researchers determined that these fungi used radioactivity as an energy source to break down food and increase growth. Studies have also shown that native deciduous trees and conifers can be inoculated with mycorrhizal fungi such as *Gomphidius glutinosus*, *Craterellus tubaeformis*, and *Laccaria amethystina*, all of which absorb radioactive cesium.

Other studies demonstrate the importance of mycorrhizal inoculation in reseeding and reclaiming forested areas damaged by mining operations. The goal is to return the mine site to natural conditions, in part by inoculating the soils with mycorrhizal fungi.

Reforestation

Reforestation efforts usually require planting commercially grown seedlings or shrubs. Commercial nurseries sometimes use sterile growing

The soil and roots of nursery-grown tree seedlings benefit from mycorrhizal fungi.

Workers replant red spruce (*Picea rubens*) in North Carolina with seedlings inoculated with ectomycorrhizal propagules.

COMMERCIAL APPLICATIONS

Many edible mushrooms, not to mention truffles, have substantial commercial value, and knowledge of ectomycorrhizal associations can be used for profitable purposes. For example, we know that if oak and hazelnut trees are inoculated with the fungus *Tuber melanosporum*, they can produce black truffles. These expensive delicacies are being grown on a commercial basis in operations from Oregon to Tasmania. Given the demand for these ectomycorrhizal fruiting bodies, this is an area ripe (sorry) for study.

media and fumigate to destroy pathogens. These conditions, however, can destroy or slow the growth of existing mycorrhizae. Although such practices may outweigh the cost benefits of fostering mycorrhizae while the plants are in the nursery, reforestation planting always benefits from the addition of mycorrhizal fungi propagules. Although ectomycorrhizal spores are light enough to be distributed by the wind from an area where the fungi are growing, the spores of arbuscular fungi are heavier and must be distributed via water, animals, or inoculation. To ensure that sufficient numbers of spores are available to the growing trees, inoculation is required.

In nurseries, inoculation of tree stock has met with mixed results—usually good, but sometimes there has been no visible difference between inoculated and uninoculated plants. It is clear, however, that when the growing conditions are not conducive to mycorrhizal health, such as when overfertilization is the rule, mycorrhizal fungi suffer.

USING INOCULANTS

The appropriate mycorrhizal association should be established as early as possible on any tree or shrub and maintained for the life of the plant. Because the appropriate fungi need to be used as inoculants, using a broad mix of mycorrhizal fungi can help ensure colonization; each fungus in the mix imparts its own particular benefit to a host plant.

Many commercial planting soils contain propagules, and liquid formulations can be applied via drip or other watering systems. Seedling roots can be dipped in granular or liquid formulations at the time of transplanting. Granular formulas can be placed in holes in the soil around drip lines, and liquid inoculants can be applied at any time to existing stock.

Growers must carefully monitor the use of fertilizers and chemicals when plants are inoculated with mycorrhizal fungi. Overfertilization, particularly with phosphorus and nitrogen, can lead to a reduction of mycorrhizal associations or even death of the fungi. Because pesticide and fungicide applications can also be detrimental to mycorrhizae, growers must apply only compatible chemicals at the appropriate times.

The amount of organic matter in the forest soil can also impact how mycorrhizae fare. In one study, mycorrhizal formation was reduced by up to 90 percent when Douglas fir seedlings were grown in soils where timber operations greatly reduced the organic material. Slash and debris should remain on site after harvest to help maintain mycorrhizal populations.

MYCORRHIZAL TREE AND SHRUB STUDIES

Thousands of studies demonstrate the efficacy of inoculating with ectomycorrhizal and, when appropriate, arbuscular mycorrhizal fungi when seeding, rooting, and transplanting shrub and tree seedlings. Here are a few summaries of the results of studies on trees and shrubs.

Almond (*Prunus* spp.) and walnut (*Juglans* spp.) One study showed that regardless of the planting medium—peat, sandy soil, or bark compost—after two years, inoculated almond and walnut seedlings showed the same amount of mycorrhizal colonization, and all inoculated seedlings grew larger than uninoculated plants. A study of the rootstocks of several almond cultivars indicated that inoculation with *Rhizophagus intraradices* resulted in larger plants and root masses. Inoculation with *R. intraradices*, *Funneliformis mosseae*, and *Claroideoglomus etunicatum* also suppressed root-knot nematodes.

Apple (*Malus* spp.) Five months after *Rhizophagus intraradices* was used to inoculate apple seedlings, the plants showed an increase in phosphorus, stem length, number of nitrogen-fixing nodules, and dry weight. When *Diversispora versiformis* and *Glomus macrocarpum* were used as inoculants, the host trees' ability to withstand drought stress was improved.

Inoculated trees grew more leaves and increased their uptake of zinc, copper, and other minerals. In addition, inoculated plants fared as well after application of low levels of phosphorus as control plants fared with higher amounts. Results were similar after inoculations of *Claroideoglomus etunicatum* and *Gigaspora margarita*.

Black walnut (*Juglans nigra*) Black walnut trees that formed arbuscular mycorrhizae with *Glomus deserticola* and *Claroideoglomus etunicatum* showed high levels of root colonization. Studies showed a strong relationship between where the seeds were harvested and which *Glomus* species was the best host. This might explain why walnuts inoculated with *Rhizophagus intraradices* produced taller seedlings than trees inoculated with *C. etunicatum*. These studies also demonstrate the importance of using the appropriate fungal symbiont for each seed source.

In addition, *Funneliformis mosseae* in soil was an effective control for stunt disease. In another study, plants inoculated with *F. mosseae* showed resistance to nematodes. Inoculated trees also showed higher amounts of several nutrients than control plants.

Chokecherry (*Prunus virginiana*) Chokecherry saplings responded to *Funneliformis mosseae* and *Rhizophagus intraradices* with greater mass above and below ground than control plants.

Douglas fir (*Pseudotsuga menziesii*) Inoculation with *Laccaria proxima* greatly improved growth of pot-grown Douglas fir. Inoculation with *Hebeloma longicaudum*, *Paxillus involutus*, and *Pisolithus arhizus* increased growth of seedlings compared to controls with increasing application of fertilizer. This suggests that if mycorrhizae are established, less fertilizer may be required.

Dwarf willow (*Salix reinii*) Dwarf willow shrubs are colonized by several species of ectomycorrhizal fungi. Improved growth and nitrogen content in seedlings grown near willows showed that the ectomycorrhizal fungi from established willows are critical to the succession of others that grow in the area.

European beech (*Fagus sylvatica*) When colonized by *Lactarius subdulcis* and *Cenococcum geophilum*, beech tree roots better withstood the impacts of drought than control plants.

Incense cedar (*Calocedrus decurrens*) Inoculation with *Rhizophagus intraradices* significantly increased the survival rates of incense cedar on a disturbed site.

Oak (*Quercus* spp.) Inoculation with ectomycorrhizal *Pisolithus arhizus* improved seedling growth, shoot diameter, and height, and inoculation with naturally occurring mycorrhizal species showed the same effect. In one study, forest soils and agricultural soils were used as media in which to grow oaks. The forest soils showed the growth of two more mycorrhizal species when compared to the agricultural soils, with better overall root and shoot growth and survival rates.

Palms (Arecaceae) Inoculations with *Funneliformis mosseae*, *Rhizophagus intraradices*, and *Glomus deserticola* were effective in increasing plant growth. *Funneliformis mosseae*–inoculated plants were hardier than uninoculated fertilized plants, suggesting that inoculating plants can result in less fertilizer use with better growth. In addition, inoculated nursery plants showed a better survival rate when transplanted to the field.

Pear (*Pyrus* spp.) Pear trees inoculated with *Rhizophagus intraradices* and *Glomus deserticola* showed increased growth, but the phosphorus levels

An inoculated pine seedling (right) shows better growth than an uninoculated seedling.

and the species affected the outcome, suggesting the importance of experimenting with inoculants to find the most effective mixture.

Pine (*Pinus* spp.) Inoculation with *Laccaria proxima* greatly improved growth of pot-grown Japanese black pine (*Pinus thunbergii*), jack pine (*P. banksiana*), and mugo pine (*P. mugo*). Inoculation with *Hebeloma longicaudum*, *Paxillus involutus*, and *Pisolithus arhizus* increased growth of seedlings compared to controls with increasing application of fertilizer. When competing with Sitka spruce (*Picea sitchensis*) in the field, inoculated pine trees were healthier after 11 years of growth. Pines with mycorrhizae also contained more nitrogen. In experiments to determine whether mycorrhizal fungi would speed and improve rooting, biologists found no significant difference between controls and the inoculated cuttings; however, when the cuttings were potted up, they showed 50 percent better growth the first year. Ponderosa pine (*Pinus ponderosa*) seedlings inoculated with *Rhizopogon roseolus* showed significantly higher survival rates when planted in the field, with a 93 percent survival rate versus 37 percent for uninoculated seedlings. Red pine (*P. resinosa*) seedlings inoculated with *H. arenosum* while still in the nursery showed greater survival rates when planted in the field.

Spruce (*Picea* spp.) Inoculation with *Laccaria proxima* improved growth of black spruce (*Picea mariana*), red spruce (*P. rubens*), and especially pot-grown white spruce (*P. glauca*). Inoculation with *Hebeloma longicaudum*, *Paxillus involutus*, and *Pisolithus arhizus* increased growth of seedlings compared to controls with increasing application of fertilizer.

Yew (*Taxus* spp.) When *Rhizophagus intraradices* was used to inoculate yew cuttings, root production was significantly higher than production in the control group, with results as good as or better than using a rooting hormone.

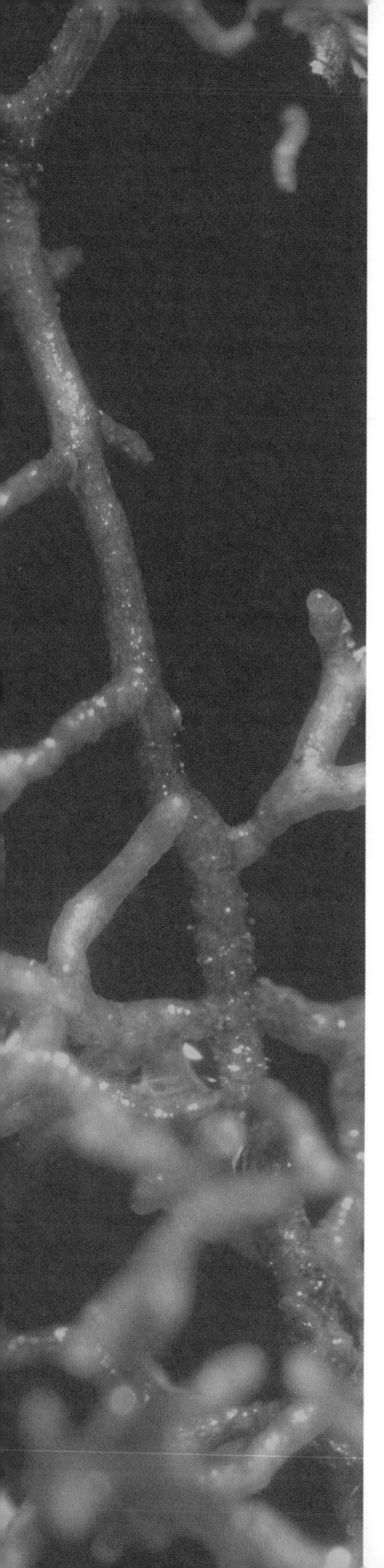

Mycorrhizae in Hydroponics

Almost any plant that can be grown in soil can be grown in a hydroponic system, from trees to fruits and vegetables. Because no soil is used, a liquid nutrient delivery system supplies all the nutrients the plants need. Why would plants also need mycorrhizal fungi? Simply put, mycorrhizal fungi offer the same benefits in hydroponic applications as they offer in planting media and soil.

Mycorrhizal fungi are good at what they do, especially in delivery of phosphorus and nitrogen to their plant symbionts. As they do in other growing systems, fungi grown in hydroponic systems expand the effective root area into the surrounding growing substrate, be it coconut fiber, rockwool, oasis cubes, gravel, vermiculite or perlite, clay pellets, or other material. The same hyphal extension that occurs with colonized roots in soil also occurs with colonized roots in hydroponic systems. Whether plants are growing in soil, a planting mix, or a hydroponic growing substrate, mycorrhizal fungi provide the same basic benefits to host plants:

- Inoculated roots in hydroponic systems are more resistant to pathogens, including *Rhizoctonia, Fusarium, Pythium,* and *Phytophthora.*
- Mycorrhizal fungi can create physical barriers around roots to protect them from disease and pathogenic fungi.
- Colonized plant roots branch and form more feeder roots, so more nutrients can be retrieved from the surrounding area.
- Mycorrhizal fungi grown in a hydroponic system can support mycorrhizospheric organisms that provide protective metabolites.

MYCORRHIZAL CONSIDERATIONS FOR HYDROPONIC SYSTEMS

As with any other growing system, for mycorrhizal fungi to thrive and benefit plants in a hydroponic system, growers must use the appropriate species of mycorrhizal fungi to colonize the plants being grown. They must also consider several other factors.

A hydroponic system that intends to support colonization of roots by mycorrhizal fungi must be fully aerobic. To ensure that the fungi will survive, growers should maintain oxygen levels at 6 to 8 parts per million (for context, most tap water provides 5 ppm of oxygen). Oxygen can be added to the growing environment via air stones or bubblers. (Although hydrogen peroxide can also oxygenate water, it will kill mycorrhizal fungi and host plant tissue and should not be used.)

Most hydroponic systems use some form of media to support plants, such as rockwool or expanded clay pebbles. Many scientists use a modified hydroponic system for their studies, using sand as a growing medium and saturating it with nutrient solution, but mycorrhizal fungi will grow in nearly any substrate, or even in water alone, if it provides enough oxygen.

Hydroponic growers often add phosphorus to nutrient delivery systems, but an increase in phosphorus levels can result in decreasing amounts of mycorrhizal colonization. When large amounts of phosphorus are present in the system, at levels of around 70 parts per million, fungal spores go dormant and will not germinate. By monitoring phosphorus levels and avoiding adding too much fertilizer, growers can encourage the formation of mycorrhizae.

The presence of chlorine and chloramines in public water systems can also affect mycorrhizal growth. Chlorine dissipates into the

TRICHODERMA AND MYCORRHIZAL FUNGI

Many growers use *Trichoderma* species in their hydroponic systems. Like mycorrhizal fungi, *Trichoderma* fungi protect plants from pathological organisms and help the associated plants increase their nutrient uptake. However, *Trichoderma* fungi generally cycle nutrients from organic matter, while mycorrhizal fungi actively transport nutrients directly into the plant cells. These two fungi work well together and both benefit hydroponic growing systems.

atmosphere in about eight hours, but chloramines take days to evaporate and should be removed from water used in hydroponic applications. Hydroponic suppliers can provide recommendations on how to accomplish this.

The pH of the water in hydroponic systems is also important. Most mycorrhizal fungi require a pH range of 5.5 to 7.0 to survive. Mycorrhizal mix packaging often lists the required pH levels.

Finally, maintaining proper temperatures will maximize mycorrhizal colonization. Arbuscular mycorrhizal fungi thrive in temperatures of 65–75°F (24–30°C), which is also the ideal temperature for most plants. Growers should be aware that vesicles, spores, and fungal hyphae will not survive temperatures above 120°F (49°C). Plants would die under such extremely high temperatures as well. When storing mycorrhizal inoculants, growers should also consider appropriate temperature conditions.

USING INOCULANTS

Hydroponic supply sources offer mycorrhizal mixes specifically designed for hydroponic systems. In these blends, spores and hyphal fragments are mixed with delivery media in liquid or powder form. To establish mycorrhizae, make sure the inoculants are in direct contact with rooted or unrooted cuttings before they are placed into the hydroponic system, or soak starter cubes in liquid formulations or granular formulations mixed with water. As in soil, the spores will germinate when they are signaled by the root exudates. Many formulations can be added directly to the hydroponic nutrient delivery system; the particles within the mix are small enough to pass through systems without clogging up lines or emitters. Mycorrhizal formulation providers should indicate the proper ways to distribute the inoculants.

Mycorrhizal fungi do not generally reproduce in hydroponic systems, so it is a good practice to add more inoculants to the nutrient delivery system as roots develop to ensure maximum colonization throughout the life of the plant. How much more and how often? The plants can offer the answers. Fast root and plant growth require more frequent addition of inoculants. Because it takes about two weeks for mycorrhizae to become established, application should be discontinued a few weeks before flowering for flowering crops and a few weeks before harvest for others, such as lettuces. Why waste propagules?

MYCORRHIZAL HYDROPONIC PLANT STUDIES

Studies show that mycorrhizal inoculation can provide many benefits to plants growing in a hydroponic system.

Cannabis (*Cannabis sativa*) *Claroideoglomus claroideum, C. etunicatum, Funneliformis geosporum, F. mosseae, Glomus microaggregatum, Rhizophagus clarus,* and *R. intraradices* are successful inoculants for cannabis growing in rockwool. Using mixes of several inoculants results in healthy plants,

Mycorrhizal fungi growing on plant roots in a hydroponic system.

but studies indicate that inoculating with *R. intraradices* alone produces similar results.

Carrot (*Daucus carota*) Carrots thrive when inoculated with mycorrhizal fungi in hydroponic systems. Nutrient sprays are best. A 3-to-1 mixture of perlite-to-vermiculite makes an ideal substrate, as long as it is deep enough to support the particular carrot cultivar. Studies showed that inoculation with *Rhizophagus intraradices* BEG 141 increased the carrots' fresh weight better than use of *Funneliformis mosseae* BEG 167.

Cucumber (*Cucumis sativus*) Cucumbers perform well when inoculated with arbuscular fungi and planted in individual cell pots with a perlite and vermiculite mix. *Claroideoglomus etunicatum, Funneliformis caledonium, F. mosseae,* and *Rhizophagus clarus* can be used as inoculants.

Garlic (*Allium sativum*) With roots that are readily colonized, garlic grows in baskets using perlite or a perlite and vermiculite mix. Cloves can be restarted in water until sprouts and roots appear and then transferred to the hydroponic system. After inoculation with *Rhizophagus fasciculatus* in field studies, garlic showed increased yields and larger bulbs.

Pepper (*Capsicum annuum*) Using netted pots and a media mixture of perlite, peat, and vermiculite, pepper plants showed increased root mass, larger fruit growth, and more accumulated nutrients after inoculation with *Rhizophagus intraradices.*

Strawberry (*Fragaria ananassa*) Many strawberry cultivars growing in a vermiculite substrate were inoculated with *Claroideoglomus etunicatum, Funneliformis mosseae,* and *Rhizophagus intraradices.* They showed better overall growth and productivity than control plants, but results varied depending on the variety.

Tomato (*Lycopersicon esculentum*) Inoculation with *Funneliformis monosporus, F. mosseae, Rhizophagus intraradices, R. vesiculiferus,* and *Glomus deserticola* increased fruit yield of tomato plants grown in a sawdust substrate. Inoculation with *F. monosporus* and *F. mosseae* increased plant height and dry weight significantly.

Mycorrhizae for Lawns and Turfgrass

Homeowners and groundskeepers alike usually fertilize their lawns and turf several times a year, adding pounds of expensive chemical fertilizers with each application. Lawn fertilizers can include high amounts of phosphorus and nitrogen, both important nutrients for grasses. But nutrient runoff has caused many pollutants to drain into rivers, lakes, and streams. In response, many American states have banned the use of phosphorus in lawn fertilizers.

The solution is arbuscular mycorrhizae. The quality of grasses grown in lawns, golf courses, and recreational fields can be enhanced by maintaining a healthy network of arbuscular mycorrhizal fungi. This practice can totally replace the application of fertilizers—in particular, the use of chemical phosphorus—because the mycorrhizal network mobilizes both phosphorus and nitrogen.

Grasses are divided into two physiological groups according to the number of carbon molecules produced during photosynthesis. Cool-season grasses are C3 grasses, and warm-season grasses are C4 grasses. The latter generally grow in tropical climates

GRASSES THAT FORM ARBUSCULAR MYCORRHIZAE

bahiagrass (*Paspalum notatum*)
Bermuda grass (*Cynodon dactylon*)
centipedegrass (*Eremochloa ophiuroides*)
Kentucky bluegrass (*Poa pratensis*)
perennial ryegrass (*Lolium perenne*)
red fescue (*Festuca rubra*)
St. Augustine grass (*Stenotaphrum secundatum*)
zoysia (*Zoysia* spp.)

and are more mycorrhizae-dependent than cool-season grasses; in fact, bahiagrass (*Paspalum notatum*), a C4 grass, is the preferred host in the production of commercial propagules. But all true grasses (Poaceae or Gramineae), including cereals, bamboo, and the grasses of lawns, golf courses, and grasslands, readily form arbuscular mycorrhizae.

Associations with arbuscular mycorrhizal fungi benefit grass plants in the same ways they benefit trees, shrubs, and most other colonized plants:

- Grasses inoculated with mycorrhizal fungi are generally healthier.
- Tests show that lawn grasses inoculated with mycorrhizal fungi contain more chlorophyll with improved photosynthesis.
- Root systems grow larger, denser, and faster as a result of mycorrhizal association, which can mean total coverage, without the need to replant.
- Mycorrhizae improve soil structure as fungal hyphae extend and explore surrounding soils. Mycorrhizae bind soil particles, allowing for better air and water movement throughout the soil.
- Mycorrhizae improve soils damaged by compaction, particularly soils supporting grass on playing fields and parks.
- Mycorrhizae improve drought resistance by reaching deep into the soil to access water resources.
- Mycorrhizae lessen infection rates from bacterial wilt, parasitic nematodes, and other pathogens.

WATER MANAGEMENT

Providing water to large expanses of grass can be expensive, time-consuming, and wasteful. Research has demonstrated that when mycorrhizae are established in lawns and turfgrass, less water is required than in areas with low colonization. In fact, in some climates, grasses with mycorrhizal associations need no irrigation. Simply put, mycorrhizae store water, but they also explore for it.

Grasses with mycorrhizal associations are also better able to resist the ravages of drought. Arbuscular fungi form tremendous extraradical networks in the soil that increase grass roots' access to water, especially in response to drought. Healthy mycorrhizal grasses with extensive root systems are better equipped to face the stresses of drought, and grasses bounce back much faster when water again becomes available. The complex mantle formed around roots by ectomycorrhizal fungi holds water, which enhances storage and the plants' ability to interact with surface water. Arbuscular mycorrhizal fungi also form abundant vesicles within grass roots that stay hydrated and spongy to protect roots from desiccation during drought.

Endomycorrhizal fungi deposit sticky glomalin into the soil, improving soil structure. Glomalin-enriched soil particles stick together, creating pores, tunnels, and reservoirs for greater water retention. Mycorrhizal colonization also increases the number of root aquaporins, the embedded cell membrane protein channels that transport water.

Mycorrhizae benefit grasses in other ways as well. For instance, they impact gas exchange in plants, helping them expel oxygen and take in carbon dioxide for photosynthesis. Mycorrhizae impact the hydraulics of the host grass plant as it balances the water absorbed through its roots and released through its stomata. Stomata pores open and close to regulate the amount of gasses and water vapor that are expelled from or held within the plant; the stomata close as water becomes scarce and open again to allow respiration.

A FEW WEEDS? WHY NOT?

In a freshly seeded lawn, the best way to discourage weed growth is to encourage the quick germination of grass seed—and mycorrhizal inoculation helps grass seeds germinate faster and establish quicker. But allowing a few weeds to grow in a lawn, golf course, or playing field is not necessarily a bad idea. For one thing, monocultures, areas planted in a single crop, are magnets for problems; if a fungal or nematode infection occurs in one place, it is likely to spread throughout a lawn planted in a single species of grass. And some weeds actually add nutrients to a lawn. Clover, for example, is an excellent addition. This nitrogen-fixing plant also forms mycorrhizae, and because mycorrhizae deliver nitrogen to grass roots, including clover in the lawn is a great way to achieve a healthy lawn without adding fertilizers. The presence of mycorrhizae also increases the number of nitrogen-fixing root nodules. Clover also holds water longer than most lawn grasses, and it shares the water with other organisms in the soil food web.

Even the dreaded yellow dandelion has its uses. It is so proficient at mining minerals with its mycorrhizal partners that it is sometimes used as a cover crop. Plantings in the following year will benefit because the soil contains more phosphorus. Because of pollution issues associated with the use of fertilizers in the United States and other parts of the world, most lawn fertilizer mixes no longer contain phosphorus. Perhaps weeds may someday take the place of the middle number in the N-P-K fertilizer trilogy.

If you'd rather not roll out the weed welcome wagon, studies show

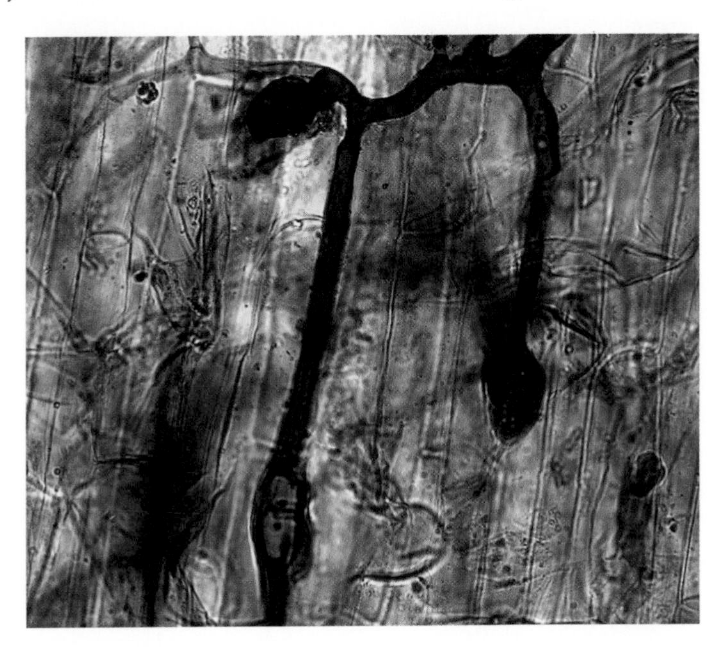

Under the microscope, mycorrhizal fungi on clover roots.

that the use of arbuscular mycorrhizal fungi can help control weed growth in established lawns and turfgrass; scientists hypothesize that the fungi either discourage the growth of the weeds or help the desired perennial grasses outcompete them.

USING INOCULANTS

Mixing spores and propagules into the soil might seem like a great way to benefit grass plants, but the fungal mix works both ways and can also give weeds a boost, because many of them form mycorrhizae as well. For this reason, inoculating lawn soils is not recommended unless a new lawn is being created on soils laid bare by recent construction. The addition of mycorrhizal fungi in this case is almost mandatory: the construction process severely damages soil structure and removes organic materials. Many of the important soil organisms that once thrived are no longer present. Mycorrhizae can help restore the soils and rebuild the food web, which benefits grasses and other plants. Golf greens benefit from the addition of mycorrhizal fungi because they are constructed mostly atop sand, which lacks organic matter and holds few nutrients; these grasses can also be bolstered by adding fungal spores. Grasses in recreational fields are exposed to all manner of stress factors, from compaction to wet and dry conditions, which can damage existing mycorrhizae. Adding propagules can replace the depleted fungi.

New lawns are usually established from seed or by laying sod. To prevent the establishment of weeds at the early stages of planting, inoculate grass seed with the appropriate mycorrhizal propagules before planting. Seed can be rolled in arbuscular mycorrhizal mixes designed for lawns, or the inoculum can be placed directly in the spreader, along with the seed, before being dropped onto the soil. If squares or rolls of sod are used, the soil side can be dusted with mycorrhizal mixes or sprayed with a liquid formula. Inoculated sod grows into the substrate faster, and the grass is more firmly established in the soil because of the extended root systems provided by the mycelial matrix.

After grass has been planted and established, fungal spores can be easily replenished; some powder mixes are small enough to wash into the root system with regular watering, but applying liquid-based inoculants can be even more efficient.

Arbuscular mycorrhizae are particularly sensitive to soil compaction

MYCORRHIZAL SPECIES FOUND IN LAWN MIXES

Funneliformis monosporus
Funneliformis mosseae
Gigaspora margarita
Glomus aggregatum
Glomus deserticola
Paraglomus brasilianum
Rhizophagus clarus
Rhizophagus irregularis

in areas where heavy traffic is the norm. The solution is aeration, which can be accomplished by hand tool or machine. If you aerate the grass prior to applying liquid or granular mycorrhizal propagules, the aeration holes offer direct routes down through the soil to the root zone.

MYCORRHIZAL GRASS STUDIES

Some grasses welcome associations with many different types of fungi; for example, in a Canadian study using soils of varying fertility, researchers found 17 species of arbuscular mycorrhizal fungi in root zones of velvet bentgrass (*Agrostis canina*) and 15 in creeping bentgrass (*A. stolonifera*).

Rhizophagus intraradices has been used to colonize Kentucky bluegrass (*Poa pratensis*), red fescue (*Festuca rubra*), and perennial ryegrass (*Lolium perenne*). Adding the less effective *Claroideoglomus etunicatum* seemed to provide growth diversity, though it was not as effective when used alone. Many other grasses, including weed grasses, have been studied for their dependency on and association with mycorrhizal fungi and for the impact these fungi can have on their growth and appearance.

Annual bluegrass (*Poa annua*) This quick-growing, opportunistic weed grass has shallow roots and is not a preferred host for mycorrhizal fungi. Experiments in golf courses have successfully used arbuscular mycorrhizae to act as a biocontrol for it; the aforementioned Canadian study found 14 species of arbuscular mycorrhizal fungi in its various root zones. Studies indicate that annual bluegrass does not thrive if mycorrhizae colonization is extensive in other, more favored, perennial grasses.

Bermuda grass (*Cynodon dactylon*) This tough grass is used in greenbelts, playing fields, and lawns. Inoculation with mycorrhizal fungi improves all aspects of growth, from drought resistance to nematode and disease resistance. Less fertilizer is also required.

Creeping bentgrass (*Agrostis stolonifera*) and Kentucky bluegrass (*Poa pratensis*) Creeping bentgrass and Kentucky bluegrass inoculated with *Funneliformis mosseae*, *Glomus aggregatum*, or *Rhizophagus intraradices* showed 60 percent higher colonization under low phosphorus conditions than controls. Creeping bentgrass showed twice as much established mycorrhizae.

Perennial ryegrass (*Lolium perenne*) After inoculation with arbuscular mycorrhizal fungi, ryegrass plants contained up to 29 percent more chlorophyll than uninoculated plants. They also showed increased biomass and better visual quality, and they contained more phosphorus, potassium, and zinc concentrations than controls. *Rhizophagus intraradices* proved to be a better colonizer than *Funneliformis mosseae*.

St. Augustine grass (*Stenotaphrum secundatum*) This grass is natively associated with *Rhizophagus intraradices*. It is susceptible to infection by brown patch (*Rhizoctonia solani*) and take-all (*Gaeumannomyces graminis*), which affect its roots. In studies, mycorrhizae did not seem to protect it from infection by these fungi.

Zoysia (*Zoysia* spp.) Zoysia grasses do not appear to be the subject of studies, but some commercial advertisements suggest that inoculation with arbuscular mycorrhizal fungi is useful and helps the grass survive drought.

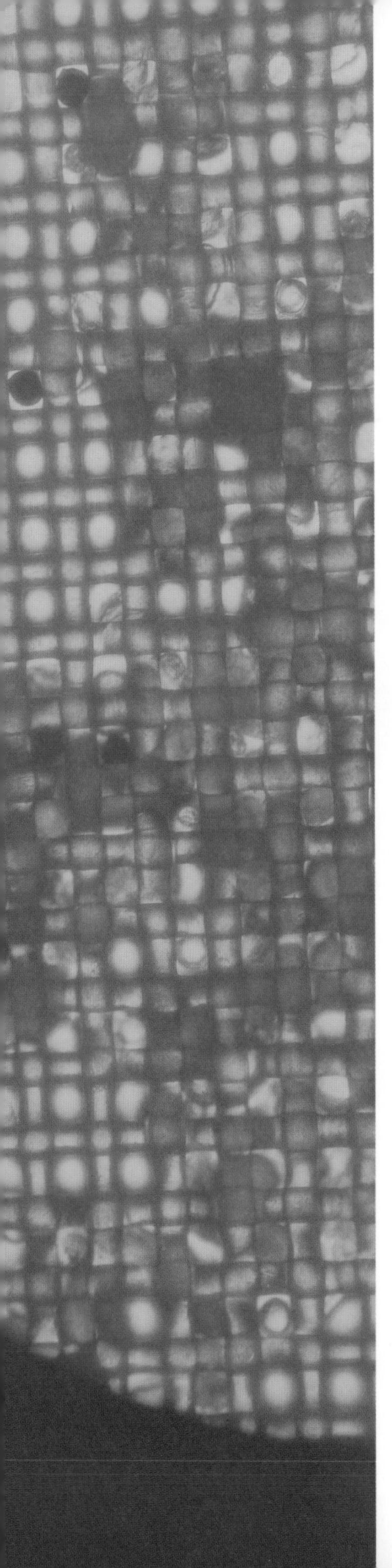

Grow Your Own Mycorrhizal Fungi

Mycologists are busy developing ways to improve methods of producing inoculum, and they are creating systems for easier reproduction of arbuscular mycorrhizal fungi. Most of these efforts stem from a desire to conserve the use of fertilizers and water for environmental and economic reasons. The result is increased availability of commercial mycorrhizal fungi products for agriculture, silviculture, horticulture, and home gardening, as well as a number of do-it-yourself systems for producing mycorrhizal fungi for inoculation.

To grow your own arbuscular mycorrhizal inoculum, you can start with spores, hyphal fragments, or colonized root fragments. Large, arbuscular mycorrhizal fungi spores, at 30 to 500 micrometers in size, can be seen with a microscope or a hand lens and can be collected easily. Moreover, these fungi can reproduce from spores or from vesicles created inside a colonized plant root. When the root dies, the vesicles germinate and develop hyphae, just as spores do.

PIONEERS OF MYCORRHIZAE IN THE LAB

For many years, although countless attempts were made to culture and grow mycorrhizal fungi in various media, they all failed. A few succeeded in growing hyphae, but they could not be independently cultured without a host plant. As obligate symbionts, however, mycorrhizal fungi cannot be grown in culture without a host root or its exudates.

In 1953, Barbara Mosse of the Rothamsted Experimental Station in Harpenden, England, achieved the first successful culture and establishment of mycorrhizae in lab conditions. She isolated and inoculated strawberry plants with a fungus now known as *Funneliformis mosseae* (named in her honor). It was also used to colonize apples, wheat, grasses, tomatoes, and lettuce, demonstrating a wide host range for a single mycorrhizal fungus.

With the ability to replicate arbuscular fungi, scientists could study them more easily. At the time, scientists were uncertain whether more than one kind of arbuscular mycorrhizal fungus existed, but by 1955, Mosse and her colleagues had identified others. And in 1961, Victoria Barrett of Alabama's Auburn University announced that she had isolated and grown arbuscular mycorrhizal fungi using hemp seed. She named the fungus *Rhizophagus*.

Several mycorrhizal fungi are named for the researchers who discovered them—Trappe, Berch, Allen, Harley, Rayner, and Schenck, for example. These scientists made discoveries of immense proportions and deserve more recognition for their efforts.

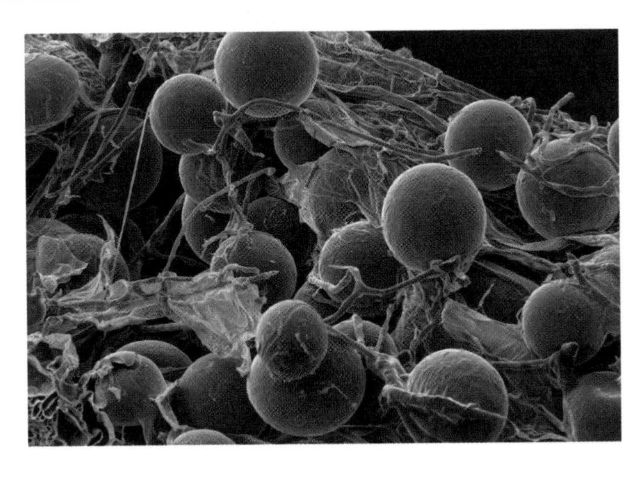

Arbuscular mycorrhizal fungal spores.

HARVESTING PROPAGULES FROM THE FIELD

You can use commercial propagule mixes with sufficient mycorrhizal fungal material and appropriate strains to match the specific crops for which the inoculum will be used. You can also gather your own propagules from soil collected from a field or from areas adjacent to a field, such as fence rows or woodlots. After you collect soils from several areas, mix them together and use a sieve to remove sticks, rocks, and other debris. This soil should contain a large and diverse population of indigenous mycorrhizal fungi to use as an inoculant; some studies suggest that indigenous mycorrhizal fungi perform better for their host plants than introduced species. If you are unsure about the soil, you can send samples to a lab to test and determine the presence and amount of propagules.

WET-SIEVING SPORES

To extract spores from soil in the lab, scientists place the soil into solution, which is then centrifuged. Spores can also be mixed with adhesive

Harvesting starter propagules from areas adjacent to productive fields produces the most diversity.

chemicals that cause them to float in a single level. This process requires laboratory chemicals to which most lay people do not have access. But even without access to or training in using lab equipment, you can use a small-to-smaller series of mesh screens to isolate the spores from the soil in solution. Relatively inexpensive kits of soil sieves are available from scientific supply companies. To do the job, you'll need a 750-micrometer mesh sieve to catch the big stuff, a 250-micrometer sieve for the next pass, then a 100-micrometer sieve, and finally a 50-micrometer sieve.

The final collection is flotation-separated: in a liquid solvent (such as water and glycerol or sucrose, also available from a scientific supply company) the bits of soil in the solvent are suspended at different levels. You can use a density gradient to measure the various densities of the

A set of soil sieves.

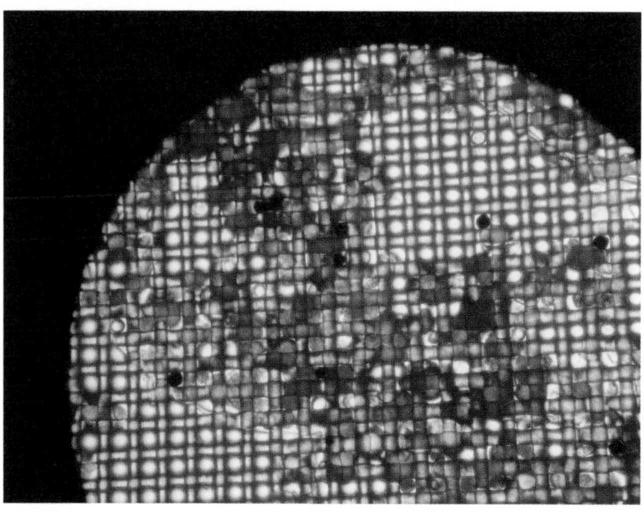

Large arbuscular spores can be caught in a sieve.

STAINING ARBUSCULAR MYCORRHIZAE FOR THE MICROSCOPE

Although the mycelium, root coverings, and fruiting bodies of some ectomycorrhizae are visible to the naked eye, it takes a dissecting microscope to view arbuscular mycorrhizae. Because the natural pigments of the plant roots are dark, obscuring the mostly transparent fungi from view, these pigments must be removed and then a stain that sticks to the fungi applied to reveal them under the lens.

To view mycorrhizal fungi in the lab, scientists wash freshly collected roots in water before bathing them in a hot solution of 10 percent potassium hydroxide (KOH) to clear out the plant parts. This is acidified with hydrochloric acid (HCl) and then finally stained.

Traditional stains are highly alkaline, which makes them too dangerous for the untrained lay person to use. Scientists take extensive safety precautions before handling them, because each poses great dangers. For this reason, the reader is not advised to use these chemicals without adequate training and supervision.

Other, slightly safer, methods can also be used for staining and viewing these small organisms. Diluted potassium hydroxide, at 2.5 percent, is a bit safer to use, though it is a hazardous material and, as such, you must be extremely careful when handling it and follow the safety guidelines on the label. Samples first need to be for soaked in water for at least 24 hours in a test tube or similar vessel. Then the samples are soaked in a 1-to-1 mixture of hydrogen peroxide-to-water. An India ink and vinegar mix works well as a stain. Further soaking in glycerol for two or three days will make the fungi even clearer.

materials. The fungal spores are heavier than most of the other organic matter in the soil, though not as heavy as minerals.

RODALE METHOD OF INOCULUM PRODUCTION

Based in Pennsylvania, the Rodale Institute has been studying organic farming and gardening methods since it was established in 1947. Its founder, J. I. Rodale, was heavily influenced by the food-growing techniques of British organic gardening pioneers Eve Balfour and Albert Howard. He wanted to promote a holistic system of agriculture that

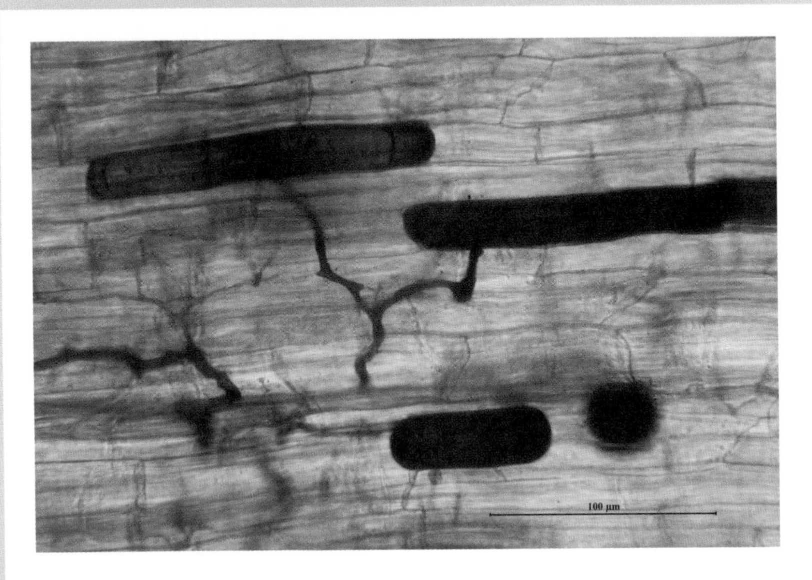

Blue-stained arbuscular fungi in a two-year-old red cedar (trypan blue dye at 0.05 percent).

As interest in growing mycorrhizal fungi increases, so will the need for simple testing procedures. Current staining techniques take too much effort and require too many skills to be conducted safely by the untrained person, however. In the meantime, biological labs will test and identify the existence and numbers of mycorrhizae for you. Identification by DNA or RNA will eventually be economical as well.

offered benefits for human health and respect for the environment. In addition to publishing several magazines and books, the Rodale Institute engages in agricultural research, partnering with private groups, colleges and universities, and the USDA.

In 2002, the institute partnered with the USDA Agricultural Research Service to develop a system that would enable farmers to produce arbuscular mycorrhizal fungi. The system needed to be simple, to use readily available and inexpensive materials, and to produce a spore-rich inoculum that was less expensive than commercial brands.

Rodale and USDA scientists spent eight years studying and developing several ways to grow arbuscular mycorrhizal fungi. In 2010, they published a description of a system they used to grow fungi that would make it more available and affordable than expensive commercial formulations. In addition, the group developed a series of recommendations with regard to standard practices that ensure the viability of the produced fungi once put to use.

The On-Farm Arbuscular Fungus Inoculum Production System was created for farmers and designed to avoid the need for complicated equipment or expensive, hard-to-find materials. (For more specific information, consult the Rodale website at rodaleinstitute.org/quick-and-easy-guide-on-farm-am-fungus-inoculum-production.)

One of the findings of the Rodale/USDA study: soil transfers run the risk of the introduction of organisms such as root pathogens. Using a host plant that is different from the crop plant prevents the inadvertent growth of pathogens associated with the final crop.

In the study, 130.8 cubic yards (100 cubic meters) of soil was spread between 16 grow bags. Rodale has experimented with the mixture used in the grow bags. Pure compost, pure sand, and perlite or vermiculite all supported mycorrhizae, but these were not successful media when used alone. They were either too low in nutrients, or, in the case of the compost, too high, particularly in phosphorus and/or nitrogen.

Compost with high nitrogen, low phosphorus, and moderate potassium levels work best. The study referenced different composts made from yard clippings, dairy manure, and leaves. The best dilution rate with the compost varied, depending on the mycorrhizal fungi used as well as the diluting media, but a 1-to-4 mixture of compost-to-diluting media worked best.

The study showed the 1-to-4 mixture of yard-clipping compost–to–vermiculite produced an average of 30 spores per cubic meter. The additional propagules come from the colonized roots, and the vesicles in them can develop mycorrhizae, too.

This method of developing inoculant can be adapted using smaller bags and different host plants. Rye (*Secale* spp.), fescue (*Festuca* spp.), corn (*Zea mays*), and other grasses will support mycorrhizae for collection of spores, propagules, and roots. With the ability to test results and identify specific fungi if need be, you can tweak the process to specific needs.

RODALE RECIPE FOR ARBUSCULAR MYCORRHIZAL INOCULUM

The Rodale recipe produces 16 bags of inoculum—enough to make 200 or 400 cubic feet (5.66 or 11.32 cubic meters) of inoculated potting media, depending on the dilution ratio of inoculum-to-media (1-to-9 or 1-to-19). The process takes four months to complete and is started in the early spring. The host plants are frost-killed and the fungal propagules overwinter in the mix right where they developed.

Materials

bahiagrass (*Paspalum notatum*) seed
field soil (or use commercial propagule mix if field soil is not available)
vermiculite
conical plastic pots
coarse-grained sand (the type used in swimming pool filters)
ground cover fabric
black 7-gallon (26.5 liter) grow bags
compost

Steps

1. Four months before the predicted last frost, germinate the bahiagrass in vermiculite.
2. Three months before the predicted last frost, transplant the seedlings into conical plastic pots filled with a 1-to-3 mixture of soil-to-sand. The conical pots provide enough space for roots to spread.
3. As soon as possible after the last frost, set up the inoculum production area by covering the ground with ground cover fabric. Then set up the grow bags. Fill the bags three-quarters full with a 1-to-4 mixture of compost-to-vermiculite. A 1-to-4 volume-based mixture of yard-clipping compost–to–vermiculite is a good start. Transplant five seedlings into each bag.
4. During the growing season, weed and water the plants in the bags as needed. Mycorrhizae will grow as the plants grow. In the winter, frost will kill the bahiagrass, and mycorrhizae will overwinter naturally outdoors in the bags.
5. The following spring, harvest the inoculum. Remove dead plant material. Shake the growing media from the root ball into a bin. This mix will contain the mycorrhizal spores and vesicle-laden hyphae. Cut the roots into short segments (less than ½ inch, or about 1 centimeter, each). Mix the inoculum into the potting media. Use a 1-to-9 mixture of inoculum-to-media for flats, or 1-to-19 mixture of inoculum-to-media for larger pots.

GREENHOUSE RECIPE FOR ARBUSCULAR MYCORRHIZAL INOCULUM

Materials

plastic pots
host plant seeds
paper towels
plastic bags
sterile growing media
commercial inoculum
low-phosphorus fertilizer

Steps

1. Sterilize plastic pots by placing them in a dishwasher with bleach-based detergent, or wash them in a solution of 1 part household bleach to 9 parts water for at least ten minutes. Then wash the pots in a dish detergent and water solution before rinsing them.
2. Soak host plant seeds for five minutes in a solution of 1 part bleach to 9 parts water. The bleach should have 5 percent available chlorine, which will be listed on the label. Alternatively, you can use 3 percent hydrogen peroxide to sterilize the seeds. Heat it to 140°F (60°C), let it cool, and then soak seed for five minutes, stirring often. In either case, rinse the seed with sterilized water after the soak.
3. Place seeds between two damp paper towels, and place these into a plastic zipper-type bag to maintain sterile conditions until the seeds start to germinate, which should occur a few days after treatment.
4. Plant seeds in 6- to 10-inch (15- to 25-centimeter) sterile pots in pathogen-free growing media, such as a 1-to-3 mixture of peat-to-vermiculite or peat-to-perlite. You can also use rockwool (presoaked), coarse-grained sand (the type used in swimming pool filters), or sterilized soil. To kill all pathogens in soil, use a mechanical soil sterilizer or place soil in a pressure cooker for

POT CULTURING INOCULUM

Inoculant can also be produced in greenhouse conditions using various host plants. Over the years, nursery growers and farmers have had success using corn (*Zea mays*), onion (*Allium cepa*), strawberry (*Fragaria* spp.), peanut (*Arachis hypogaea*), sorghum (*Sorghum* spp.), and big bluestem (*Andropogon gerardii*) as hosts. Warm-weather plants can also be about an hour until the soil reaches 140–158°F (60–70°C). The media must not contain phosphorus because this will prevent formation of sufficient mycorrhizae.

5. Thoroughly mix the commercial inoculum into the sterilized media, 1 part inoculum to 20 parts growing media. Roll the seeds in the mix or sprinkle the mix over the seeds. Plant three to six seeds per pot. (Note: instead of using commercial inoculum, some growers use spores from soil known to contain them, diluted soil that contains spores, or soil taken from inoculated fields; others use transplants with established mycorrhizae.)
6. Feed host plants periodically; they are in sterile growing media without working microbes. General chemical fertilizers suggest feeding every ten to 12 days, but this must be adjusted for each type of plant. Use a fertilizer that will not discourage mycorrhizae formation, one with a low phosphorus number (the P in N-P-K) of 6 or less. Some examples are alfalfa meal (2-1-3), fish emulsion (5-2-2), and soybean meal (7-2-1). You can also use compost, compost teas, or diluted fertilizer mixes.
7. After about six weeks, mycorrhizae will begin to form. This is a critical period. At any time, you can take samples of roots and have them tested to determine colonization rates.
8. After about 14 weeks, stop watering to let the planting media dry out.
9. At around week 16, gently pull the plants out of the media, being careful not to damage the roots. Shake off the media surrounding the roots, which contains spores and fungal fragments, and store it in a container in a cool and dry location. Cut the roots into small pieces; these contain vesicles and fungal hyphae. Store them in a little of the media in a cool and dry location. These propagules will remain viable for about a year.

used as host plants, including pencilflower (*Stylosanthes* spp.), bahiagrass, and kudzu (*Pueraria phaseoloides*).

FIELD CULTURING INOCULUM

To grow your own mycorrhizal inoculum, you can start with any commercial soil mix that does not include added fertilizers or manures.

Sterilize the commercial mix in an autoclave or oven for about an hour and a half at 260°F (121°C). (Sterilizing soil in an oven can create unpleasant odors.) After a week or two, mix 1 part field-collected soil with 3 parts sterilized soil, or use commercially available mixes of fungi instead of field soil to prevent pathogens.

The best soil and propagule collection site is a field that has been in production previously or is presently in production and that offers soil rich in organic material. The better the crop grown in the soil added to the mix, the more likely crops inoculated with the propagule mix will thrive. In addition, the higher the diversity of the field soil's mycorrhizae, the better. You can have the field soil tested for propagules before using it.

You can also gather propagules from strawberry plants using field culture. First, transplant a strawberry plant from the field to a pot filled with sterile soil. Or establish a runner in potted field soil and transplant it into a pot with sterile soil. Nurture plants until they are mature and well established. Then remove the plants and roots, save the soil clinging to the roots, and chop up the roots. The resulting products can be used to inoculate plants started in sterile soil.

These methods would likely be viable for use with other crops as well. Experiment to find out.

PRODUCTION OF ECTOMYCORRHIZAL FUNGI

Ectomycorrhizal fungi are easy to produce provided you have an appropriate host plant. Amateurs and professional foresters and landscapers can help plants develop ectomycorrhizae using several methods.

Studies have shown that soil containing ectomycorrhizal fungi taken from a forest at one location can be successfully used to associate with host trees at another. One such study used soils taken from New Zealand, India, Italy, Wisconsin, and the Alaskan tundra to establish successful mycorrhizal associations with the same tree hosts in one plot. Conifers and deciduous trees respond equally to these treatments.

Collect the fruiting bodies (such as mushrooms) of ectomycorrhizal fungi to use as propagules. This takes some preparation and a bit of knowledge—based on the season and conditions, the appearance and thus availability of fruiting bodies can vary. The sporocarps are loaded with spores. Experienced mushroom hunters know how to make a spore

FOUR WAYS TO HARVEST ARBUSCULAR FUNGAL SPORES FROM THE FIELD.

Spores isolated from field soil

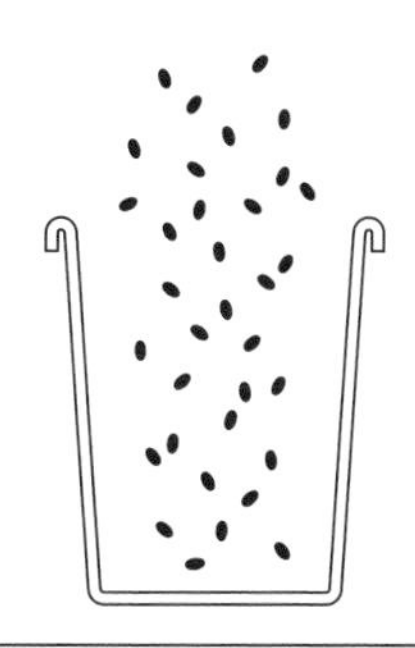

Culture quality
usually single species
Success rate
moderate

Spores in diluted field soil

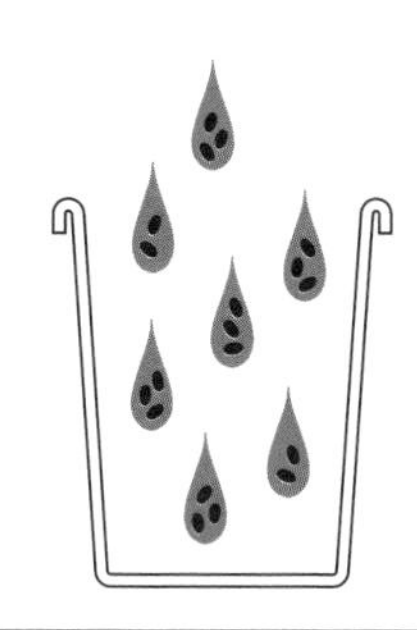

Culture quality
usually mixed species
Success rate
high

Root fragments from field soil

Culture quality
mixed or single species
Success rate
high

Seedling transplanted from field soil

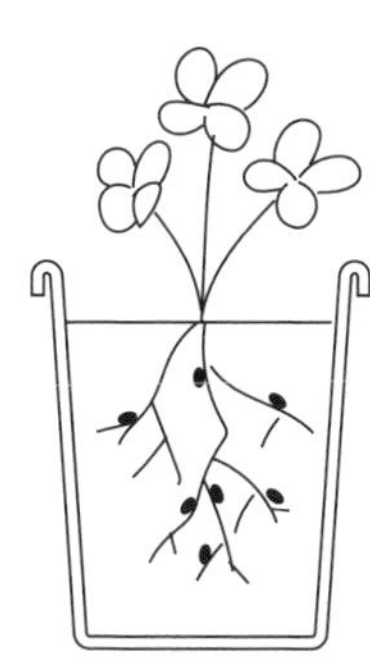

Culture quality
often single species
Success rate
moderate

FIELD CULTURE VERSUS POT CULTURE

Comparisons over the years have demonstrated that pot culturing arbuscular mycorrhizal fungi is usually far more successful than field culturing. The overcrowding of the fungal hyphae in the pot appears to encourage and increase spore production. The fungi and roots grow to the point at which the pot constricts the growth of the fungi and they can no longer expand by adding hyphal branches. The fungi are still receiving carbon, however, so they channel that energy into sporulation.

print to identify a mushroom by removing the stem and placing the cap, spore side down, on a piece of paper or glass for 24 hours. Spores will be deposited onto the paper or glass and can be collected.

To establish ectomycorrhizal colonies, add propagules directly adjacent to plant roots or spray seeds with spore suspensions. For the home gardener and small grower, a much easier method is to cut the fruiting bodies into small pieces and mix them into potting soil. You don't need more than a few spores to inoculate the soil, and it is impossible to use too much. This is a great way to grow a single strain or to make your own mix.

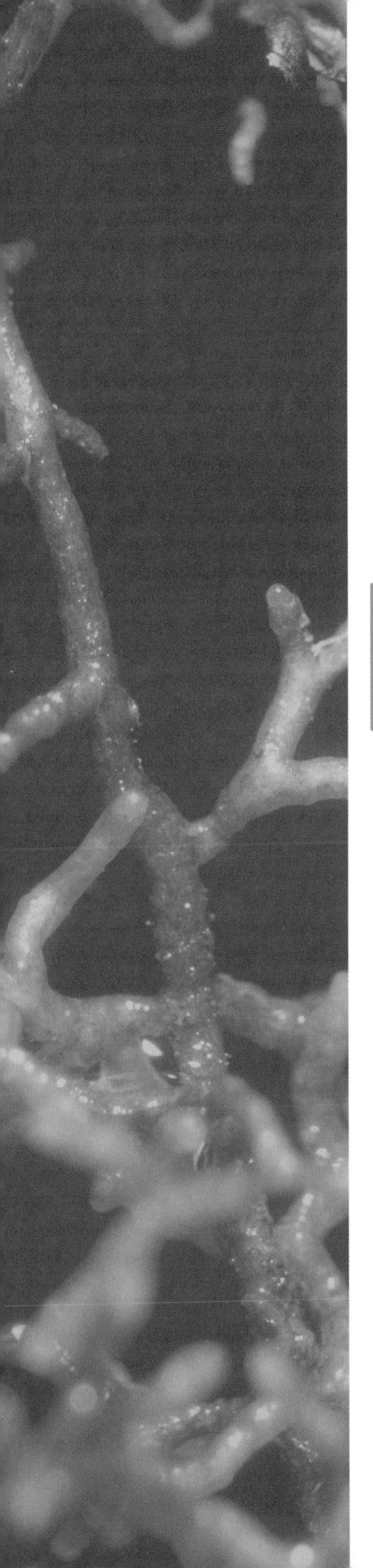

Mycorrhizae Rule!

Mycorrhizal fungi are clearly important and valuable tools for every grower: farmer, horticulturist, landscaper, turf specialist, silviculturist, or home gardener. Every growing environment is different. Mycorrhizae can be temperamental, and a mycorrhizal pairing may work in the greenhouse but not in the field. Different conditions within an environment can produce different results. You will need to experiment to learn what works best. There are, however, some general rules you can follow for the best results.

Match the fungi to the plant

Mycorrhizal associations are host-specific. Although most trees associate with ectomycorrhizal fungi and flowers and shrubs associate with arbuscular endomycorrhizal fungi, some variability occurs. Using the right fungus or fungi can be important to obtain the best results. Benefits of mycorrhizae depend on conditions but can also vary according to the particular species of fungus and host plant.

Using a mixture of various species of mycorrhizal propagules often results in the greatest benefits and usually increases the probability of your getting a good match. Each fungal species may impart its own unique abilities to offer the most benefits in total.

Not only is the proper mix important, but the delivery system makes a difference. Liquid formulations may be the best application for lawns, but they may not be appropriate for tree seedlings or some farm crops. Read labels carefully for advice regarding the best plant associations and to determine which particular species of propagules are included in the mix.

Use viable propagules

Make sure you are using viable propagules. Propagules include spores, fungal hyphae fragments, and root fragments with vesicles, but some commercial mix labels list only spore counts. Commercially produced spores last for a couple of years, and root fragments with vesicles are viable for about a year. Although they won't all die at once, the number of viable propagules decreases with time. You can inoculate and then test the soil for their presence, or you can send propagules to a lab to

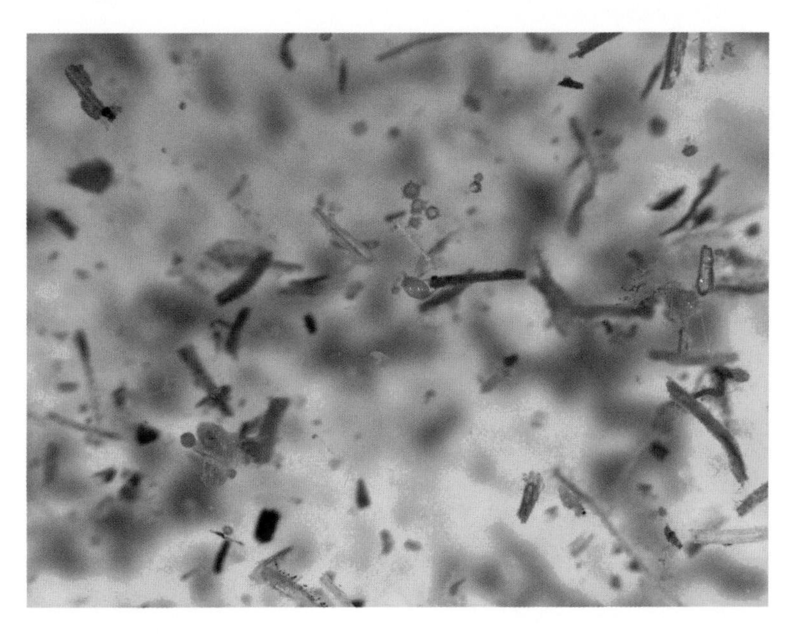

Propagules include spores (center) and root fragments that contain vesicles.

test them for viability. The manufacturers of these products may also provide batch test viability results.

Watch the phosphorus

If the soil contains too much phosphorus, a potential host plant will not create, partner with, and form mycorrhizae. To ensure continuous colonization, avoid the use of fertilizers that contain phosphorus. If you send soil to a lab for analysis, look for phosphorus levels of less than 70 parts per million. Some suggest that phosphorus levels should be lower, in the range of 30 parts per million.

Add propagules to soil mixes

Add mycorrhizal propagules to soil mixes, including compost mixes, and use these when you're transplanting, even if plants have already been colonized. Tests show that early colonization maximizes benefits, but colonies do not necessarily transfer to new roots in a new container if they are not replenished. The more propagules there are, the greater the chances that fungal hyphae will encounter roots.

Inoculate as early as possible

Give the plant a head start so that the benefits of mycorrhizae are available immediately and continuously from germination. To start the formation of mycorrhizae as early as possible, roll seed in propagules. When you purchase a plant, add propagules to the soil and/or roots before you transplant.

Beware –icides

Pesticides, herbicides, and fungicides impact both the mycorrhizal fungi and the soil organisms that interact with them. Adding the wrong –icide can kill mycorrhizal fungi. Consult the product label or online information to determine how a substance may impact precious fungi. On the other hand, you may be surprised to see which chemicals aid mycorrhizal colonization.

Be gentle

Because mycorrhizal fungi and networks are fragile, you must treat them with care. Disturb soils as little as possible, and avoid tilling, deep plowing, leveling, or other mycelium-destroying activities. Do not burn fields to remove stubble; high temperatures can kill the arbuscular mycorrhizal fungi and spores. Because fungi need oxygen to survive, avoid practices that result in soil compaction and poor drainage. If you add propagules to compost teas, add them immediately before application.

Store propagules properly

Store propagules within a range of temperatures that will keep them alive: 65–75°F (24–30°C) is best. Temperatures above 120°F (49°C) will damage or destroy propagules. Keep them cool, dry, and contained to avoid contamination. If you purchase and use a propagule mix, consider transferring the package to a clean, sealable glass jar; if the package is too large to fit inside, sterilize the jar and pour the mix inside. Then store the tightly sealed container in a dark, cool location, and keep the label on hand so you won't forget what's inside and so you can periodically check the viability.

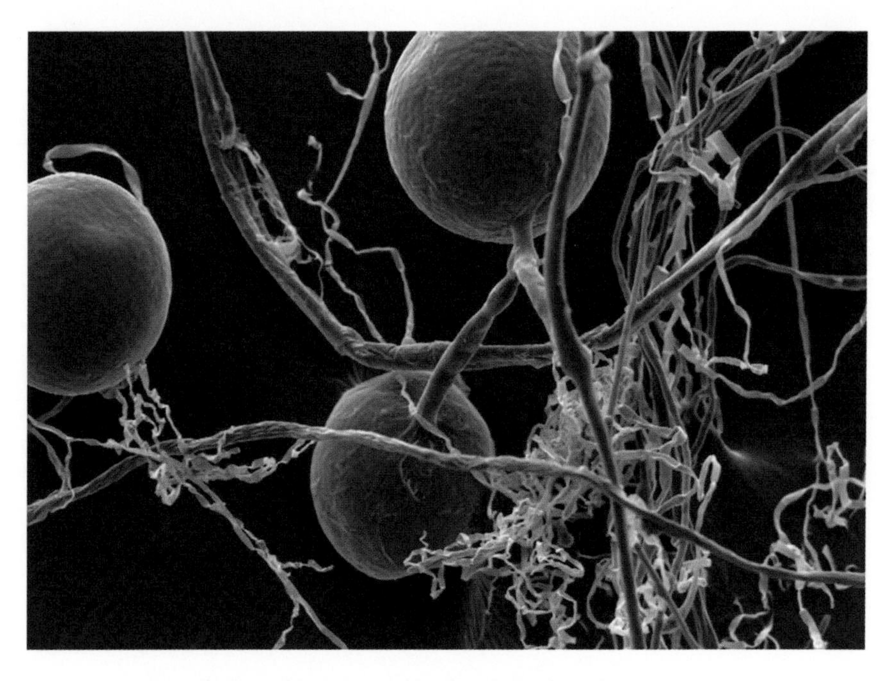

A view of arbuscular spores and hyphae through an electron microscope shows the fragility of these organisms.

Stay up-to-date

Although mycorrhizae were discovered more than a century ago, their full potential is still not known. A tremendous amount of research continues, as scientists look for ways to use mycorrhizae to increase production of food and other crops. Many universities conduct mycorrhizal research, and new studies are published regularly and available via the Internet. Several sites offer abstracts of scientific papers related to mycorrhizal fungi.

Experiment

Using propagules to create mycorrhizae is a relatively new process for growers and begs for citizen science. Experiment to find ways to optimize existing processes and improve your results. Remember that results vary, even when scientists study the fungi. For example, a USDA study reviewed 568 combinations using 66 host plants and 74 fungal species. Of these combinations, 267 showed positive results and 253 showed no impact. Some studies showed that inoculated plants developed more mycorrhizal associations when no fertilizer was used, but in others, associations were best with the addition of fertilizer. Some host plants establish more mycorrhizal associations when bare roots are planted, but others do better with inoculation of the soil in which they are grown.

We still have a lot to learn. Experiment to find the best fungal combinations for the plants in your soil. And then share your results.

Mycorrhizae and the Future

A dedicated group of mycologists has made a tremendous amount of progress in advancing the understanding and importance of mycorrhizae. These scientists have demonstrated that mycorrhizae are not only critical to life on Earth, but that mycorrhizal fungi are beautiful and mind-blowing in their function and form. What was once thought to be a lone and supposedly pathogenic fungus has become a multitude of species bearing nothing but benefits.

Continuing discoveries that contribute to our knowledge of mycorrhizal fungi offer the promise of an interesting, even brighter, future. Given the many problems we've created and now face with soil degradation and erosion, not to mention reduction in phosphorus production as reserves are depleted, we'd be crazy not to welcome the relief mycorrhizae offer with open arms. And with the many ecological impacts of mycorrhizae, we can solve problems biologically rather than by using chemical solutions with deleterious results. As mycorrhizal propagules become less expensive to produce, with better viability and delivery systems, they are becoming economically practical for every application—from forest, to cropland, to nursery, to home garden.

Studies have revealed specific protein transporters that enhance the ability of mycorrhizal fungi to absorb heavy metals, including radioactive toxins. They sequester these toxins within their cells, or the host plant can use carboxylic acids, such as citric, malic, and malonic acids, to sequester heavy metals in leaf vacuoles. Reforestation efforts following tree harvest, fires, or removal of minerals can also benefit from the reintroduction of mycorrhizal propagules into forest soils. In nutrient-poor and water-deficient forest environments, mycorrhizal fungi play an important role in helping recently planted trees absorb more nutrients and moisture from stressed soils, increasing growth and improving overall forest health.

Biochemical and molecular studies may prove helpful for genetically improving mycorrhizal fungi that can be adapted to handle a variety of problems associated with plant health in the forest as well as the field. As warming temperatures around the globe impact agricultural production and water availability, mycorrhizal fungi can be useful in introducing new plants to new climate zones or in maintaining existing plants in changing zones, helping them to adapt. In addition, the impacts of drought can be mitigated by mycorrhizae that extend to access underground reserves of water that plant roots alone cannot reach.

The glomalin glycoprotein created by arbuscular mycorrhizal fungi may act as an important carbon sink in productive soils. As some of the carbon dioxide consumed by plants is transferred to their mycorrhizal fungi as glucose from photosynthesis, much of it is eventually turned into glomalin, which is stored in the soil for many years before it biodegrades. The sticky glomalin also improves soil structure and reduces soil erosion and compaction by helping soil form aggregates to improve soil porosity. These soils can store more air and water to benefit healthy growth of roots and important soil microbes.

We are better able to identify mycorrhizal fungi using DNA and RNA sequencing, giving us a leg up in identifying fungal varieties and their associations with particular host plants. Work in the field of fungal genetics is leading to the creation of tailor-made, perhaps even condition-specific, mycorrhizae. As new techniques improve spore production and results are demonstrated in the field, use of mycorrhizal fungi will become the norm.

We still have a tremendous amount to learn about mycorrhizal fungi and mycorrhizae. We know that other organisms work with mycorrhizal fungi in the mycorrhizosphere, but we have yet to identify them all and understand exactly how they work. When A. B. Frank identified mycorrhizae and explained their role some 150 years ago, he could only guess at their importance. Now that we know, it is time to use this knowledge.

Resources

Supplies and Tools

ASC Scientific
ascscientific.com/sieves.html#tyler
Sieves for spore identification and collection.

Stuewe and Sons
stuewe.com/products/rayleach.php
Conical plastic pots.

Sources of Mycorrhizal Fungi

Australia
MicrobeSmart
microbesmart.com.au

Canada
Premier Tech
premiertech.com/global/en/products/horticulture-agriculture

Czech Republic
Symbiom
symbiom.cz

France
INOCULUMplus
inoculumplus.eu

Germany
Wilhelms Best
wilhelmsbest.de

Greece
Farma-Chem
farmachem.gr

Hungary
Agro Bio
agrobio.hu

Italy
Mybatec
mybatec.eu

Mexico
Plant Health Care
phcmexico.com.mx

Netherlands
Servaplant
servaplant.nl

Slovenia
Symbiom
mikoriza.si

South Africa
Biocult
biocult.org

South Korea
Top Blueberry
topblueberry.co.kr

United Kingdom
The Nutrient Company
thenutrientcompany.com

PlantWorks Limited
rootgrow.co.uk

Symbio
symbio.co.uk

United States
Fungi Perfecti
fungi.com

Mycorrhizal Applications
mycorrhizae.com

Plant Health Care
planthealthcare.com

Reforestation Technologies International
reforest.com

Santiam Organics
santiamorganics.com

Valent BioSciences Corporation
valent.com

ViaTerra
viaterrallc.com/#!mycorrhiza

Testing Laboratories

Australia
Soil Foodweb Institute
soilfoodweb.com.au

Canada
Soil Foodweb Canada
soilfoodweb.ca

New Zealand
Soil Foodweb Institute New Zealand
soilfoodweb.co.nz

United States
Earthfort
earthfort.com

Harrington's Soil Testing Laboratory
harringtonsorganic.com

Microbial Matrix Systems
microbialmatrix.com

MycoRoots
mycoroots.com

Mycorrhiza Biotech
mycorrhizabiotech.com

Mycorrhizal Applications
mycorrhizae.com/customer-support/support/lab-forms

ROI Microbiotics Testing Laboratory
thesoilguy.com/SG/ROITestingLaboratory

Soil Foodweb New York
soilfoodwebnewyork.com

Ward Laboratories
producers.wardlab.com

Western Laboratories
westernlaboratories.com

Further Reading

Baker, Linda. 2002. "How mushrooms will save the world." Salon.com, 25 November 2002. salon.com/2002/11/25/mushrooms.

Fungi Perfecti. 2016. "About Paul Stamets." fungi.com/about-paul-stamets.html#sthash.8QRYCeKY.dpuf.

Harley, J. H., and E. L. Harley. 1987. "A checklist of mycorrhiza in the British flora." *New Phytologist* 107(4): 741–749.

Isaacson, Andy. 2009. "Return of the fungi." *Mother Jones*, November/December 2009. motherjones.com/environment/2009/11/paul-stamets-mushroom.

Miyasaka, S. C., et al. 2003. "Manual on arbuscular mycorrhizal fungus production and inoculation techniques." *Soil and Crop Management*, July 2003. www2.ctahr.hawaii.edu/oc/freepubs/pdf/SCM-5.pdf.

Olson, Michael. 2007. Interview with Paul Stamets. *The Food Chain* (radio program), January 2007. metrofarm.com/assets/podcasts/2007-01-07_525dmushroom.mp3.

Phelps, Megan. 2009. "Why we need mushrooms." *Mother Earth News*, 13 January 2009.

Stamets, Paul. 2005. *Mycelium Running: How Mushrooms Can Help Save the World*. Berkeley: Ten Speed Press.

——. 2011. "How mushrooms can clean up radioactive contamination—an 8 step plan." *Permaculture*. permaculture.co.uk/articles/how-mushrooms-can-clean-radioactive-contamination-8-step-plan.

St. John, Ted. 2000. *The Instant Expert Guide to Mycorrhiza*. green-diamond-biological.com/wp-content/uploads/2012/03/Mycorrhiza-Primer.pdf.

Wierschem, Jenny. 2007. "Fungi to the rescue." *Forest Magazine*, Winter 2007.

Publications Devoted to Mycology

Field Mycology (britmycolsoc.org.uk/society/publications/journals/field-mycology). A journal published by the British Mycological Society, is devoted to the identification and study of wild fungi in Britain and Europe.

Fungal Biology (britmycolsoc.org.uk/society/publications/journals/fungal-biology). The international research journal of the British

Mycological Society. It publishes original contributions in all fields of basic and applied research involving fungi and funguslike organisms.

Mycological Research (journals.cambridge.org/action/displayJournal?jid=MYC). Published for the British Mycological Society, this is the leading international monthly for original research on all aspects of fungi (including lichens, slime molds, and yeasts): molecular biology, physiology, plant pathology, systematics, ultrastructure, biochemistry, biodeterioration, biotechnology, and genetics.

Mycology: An International Journal on Fungal Biology (tandfonline.com/toc/tmyc20/current). The official journal of the Mycological Society of China.

Mycorrhiza (springer.com/life+sciences/microbiology/journal/572). An international journal published by Springer, which covers research into mycorrhizae, including molecular biology, fungal systematics, development and structure, and effects on plants.

New Phytologist (newphytologist.org/journal/about). A publication of the New Phytologist Trust, a not-for-profit organization dedicated to the promotion of plant science, includes high-quality, original research.

Websites Devoted to Mycology

Many sites offer great information about mycorrhizal fungi. I maintain a list of websites on the systematics pages at MykoWeb: mykoweb.com/systematics/index.html.

DEEMY: An Information System for Characterization and Determination of Ectomycorrhizae (deemy.de). A huge database with descriptions and photographs.

Dictionary of the Fungi (speciesfungorum.org/names/fundic.asp). Lots of fungal classification information here.

Fungi Perfecti (fungi.com). Dedicated to promoting the cultivation of high-quality gourmet and medicinal mushrooms.

GINCO, Glomeromycetes in vitro collection (agr.gc.ca/eng/science-and-innovation/research-centres/ontario/eastern-cereal-and-oilseed-research-centre/the-glomeromycetes-in-vitro-collection/?id=1236786816381). Agriculture and Agri-Food Canada site offers lots of information about mycorrhizae. You can also learn about how AAFC provides the scientific community and industry with a source of high-quality starting inoculum of arbuscular fungi, produced under in vitro conditions and free of contaminants.

International Bank for the Glomeromycota (i-beg.eu). A great database of spore images.

International Culture Collection of (Vesicular) Arbuscular Mycorrhizal Fungi (invam.wvu.edu). Information from West Virginia University on collection, cultures, methods, and fungi.

Mycorrhizal Applications (mycorrhizae.com). A comprehensive website on all things mycorrhizal fungi. Great lists of hosts, studies, pictures, practical guides, and more. These folks were early pioneers in the production of mycorrhizal fungi and are the world's largest producers of mycorrhizal inoculants.

Mycorrhizal Associations: The Web Resource (mycorrhizas.info). An encyclopedia of mycorrhizae information. Some of the links are defunct, but the diagrams, information, and photos are all helpful.

Mycorrhiza Literature Exchange (mycorrhiza.ag.utk.edu). As the name suggests, this site provides an exchange of articles about mycorrhizae. Abstracts galore!

Mycorrhiza Network at TERI (mycorrhizae.org.in). News, literature abstracts, and meeting announcements.

MycorWeb (mycor.nancy.inra.fr/index.html). Research initiative aimed at deciphering the complex tree–fungus–bacteria systems. In particular, it draws on programs in comparative and functional genomics, biochemistry, bioinformatics, and ecology.

Plant Success (plant-success.com). Producers of a line of retail products for homeowners and hydroponic growers, with handy guides to liquids, powders, granular formulations, and tablets for easy inoculation.

Reforestation Technologies International (rti-ag.com). Consultants and scientists, producers of *Rhizophagus irregularis*. A useful site with good products.

Vilgalys Mycology Lab, Duke University (sites.duke.edu/vilgalyslab/resources). Great site full of links, including rDNA primer sequences, truffle information, and more.

Photo and Illustration Credits

ILLUSTRATIONS

All illustrations are by Winnifred S. Casacop, with the exception of those on pages 19, 35, 65, 69, 112, and 145, which are by Arthur Mount.

PHOTOS

M. **Amaranthus,** Mycorrhizal Applications, pages 33, 45, 50, 56, 57, 68, 75 (courtesy Jim Deacon, University of Edinburgh), 78, 79, 92, 93 (courtesy Eddie Loy, Star Seed), 95, 96, 97, 103, 104, 110, 120, 135, 137 (bottom), 148, and 150.
Lynn **Betts,** USDA-NRCS, page 77.
Efren **Cazares,** MycoRoots, pages 87 and 111.
Michelle M. **Cram**, USFS, page 139.
Nigel **Davenport,** The Nutrient Company, page 125.
Patricia D. **Duncan,** Environmental Protection Agency, page 91 (courtesy National Archives).
Kristine **Nichols,** USDA-ARS, page 49.
Charles **O'Rear,** page 82 (courtesy Special Media Archives Services Division, NWCS-S).
USDA-ARS, page 17.
USFS, Northeastern Area Archive, page 59.

Bugwood.org

Used under a Creative Commons Attribution 3.0 license
Robert L. **Anderson**, USFS, page 60 (top).
Joseph **O'Brien,** USFS, page 107.

Used under a Creative Commons Attribution–Noncommercial 3.0 license
J. H. **O'Bannon,** page 74.

Flickr

Used under a Creative Commons Attribution 2.0 Generic license
SuSanA Secretariat, page 136.

Wikimedia

Used under a Creative Commons Attribution–Share Alike 4.0 International license
Rowan **Adams,** page 72.
Mylène **Durant,** page 130.

Used under a Creative Commons Attribution–Share Alike 3.0 Unported license
BMK Wikimedia, page 137 (top).
Roland **Gromes,** page 15.
André-Ph.D. **Picard,** page 44.
Silk666, page 116 (top).

Used under a Creative Commons Attribution 3.0 Unported license
H. **Krisp,** pages 39 (top) and 60 (bottom).
Forest and Kim **Starr,** hear.org/starr, page 109.

Used under GFDL and Creative Commons Attribution 3.0 Unported licenses
Mike **Guether,** page 48.

Used under a Creative Commons Attribution–Share Alike 2.5 license
John **Womack,** page 27.

Used under a Creative Commons Attribution–Share Alike 2.5 Generic license
Bernd **Haynold**, page 63.

Used under a Creative Commons Attribution–2.5 Generic license
David P. **Hughes** (top), Maj-Britt Pontoppidan (bottom), page 28.

Used under a Creative Commons Attribution–2.0 Generic license
U.S. Fish and Wildlife Service Southeast Region, page 116 (bottom).

Used under a Creative Commons Attribution–Share Alike 2.0 Germany license
Jörg **Hempel,** page 39 (bottom).

Released into the Public Domain
George **Chernilevsky,** page 113.
Masur, page 14.
Jonson22, page 80.

Index